MODELING AND SIMULATION: THEORY AND PRACTICE

A Memorial Volume for Professor Walter J. Karplus

MODELING AND SIMULATION: THEORY AND PRACTICE

A Memorial Volume for Professor Walter J. Karplus (1927-2001)

EDITED BY

GEORGE A. BEKEY
University of Southern California,
Los Angeles, California

BORIS Y. KOGAN
University of California
Los Angeles, California

KLUWER ACADEMIC PUBLISHERS
Boston/London/Dordrecht

Distributors for North, Central and South America:
Kluwer Academic Publishers
101 Philip Drive
Assinippi Park
Norwell, Massachusetts 02061 USA
Telephone (781) 871-6600
Fax (781) 681-9045
E-Mail: kluwer@wkap.com

Distributors for all other countries:
Kluwer Academic Publishers Group
Post Office Box 322
3300 AH Dordrecht, THE NETHERLANDS
Telephone 31 786 576 000
Fax 31 786 576 474
E-Mail: services@wkap.nl

Electronic Services < http://www.wkap.nl>

Library of Congress Cataloging-in-Publication Data

Modeling and simulation : a memorial volume for Professor Walter J. Karplus (1927-2001) / edited by George A. Bekey, Boris Y. Kogan.
p. cm.
Includes bibliographical references and index.
ISBN 1-4020-7062-4 (alk. paper)
1. Computer simulation. 2. Karplus, Walter J. I. Bekey, George A., 1928- II. Kogan, Boris Y. (Boris Yakob), 1914-

QA76.9.C65M6179 2003
003.3--dc21

2002043388

Printed on acid-free paper.
Printed in the United States of America

Life Is Transient But
Knowledge Is Eternal

Walter J. Karplus

1927 - 2001

Walter J. Karplus, one of pioneers in the field of computer science, was born in Vienna, Austria, on April 23, 1927. He became a U.S. citizen in 1944 and served in the U.S. Navy from 1945 to 1946, before earning a B.S. degree at Cornell University in 1949. He received a master's degree at U.C. Berkeley in 1951 and a PhD at U.C.L.A. in 1955. Dr. Karplus joined the U.C.L.A. faculty in 1955 and was one of the founders of its Computer Science Department. From 1972 to 1979 he served as the Department Chair. He was also head of the Computer Simulation Laboratory and served for many years as Director of U.C.L.A.'s Center for Experimental Computer Science. At the time of his death from cancer on November 11, 2001, Dr. Karplus was Interim Dean of U.C.L.A.'s Henry Samueli School of Engineering and Applied Science.

A registered engineer in California and the holder of four patents, Dr. Karplus held scientific positions at various times with Hughes Aircraft Co., International Geophysics and Sun Oil. He was a frequent consultant for governmental agencies and industrial organizations and often served as an expert witness about computer-related matters. He served as a Director of the American Federation of Information Processing Societies. In recognition of his work on human-computer interfaces, virtual reality and neural networks, Dr. Karplus was elected President of the IEEE Neural Networks Council. His many awards include a Fulbright Fellowship, a Guggenheim Fellowship, the Senior Scientific Award of the Society for Computer Simulation, the Silver Core Award of the International Federation for Information Processing and an Achievement Award from NASA.

Dr. Karplus was the author or co-author of nine books, and 130 technical papers on a broad range of subjects. His first book, published in 1958, was the pioneering work on simulation and the solution of field problems and he continued to make significant contributions to this field throughout his career. For example, in the 1990s his research successfully employed virtual reality to allow physicians to observe how blood flows through aneurysms deep inside the human brain.

Dr. Karplus is survived by his wife, Takako, their two children, Maya and Anthony, and his brother, Curt.

Preface

This book is dedicated to the memory of a dear friend and respected colleague, Prof. Walter J. Karplus from UCLA. As the title of the book indicates, Dr. Karplus was a pioneer in modeling and simulation, and many of the chapters have resulted from collaboration with him or from inspiration resulting from his work.

Hence, this book should not be viewed as an epitaph, but rather as a celebration of this man's personal and professional achievements. Our purpose is not to live forever, but to live well and fully and to contribute to society to the best of our ability. This he has done, in full measure,

Walter, we miss you, but the work you started will be continued and expanded, not only by ourselves, but by your numerous other colleagues throughout the world. This is your legacy to the profession.

George A. Bekey and Boris Y. Kogan

CONTENTS

PART II
Methodology

PART III
Applications of Simulation

Introduction

Our friend and colleague, Professor Walter Karplus from the UCLA Computer Science Department, died of cancer in November 2001 at the age of 74. He was a pioneering contributor to the field of modeling and simulation, having published numerous books and articles in the field for nearly 50 years. In honor of his many contributions we have assembled this volume of articles written by former students, collaborators, and colleagues whose work was influenced by Walter's research.

It is significant that the papers in this book cover a wide range of topics, since Prof. Karplus had wide interests and worked in a multitude of areas where modeling and simulation could be applied.

The book is divided into three parts. Part I contains a list of all of Professors Karplus's former Ph.D. students and a list of all his publications. Clearly, his students and papers are his most tangible legacy. Chapter 1 summarizes his technical and scientific contributions. The following four chapters present brief reminiscences from colleagues, friends, and a former student. They provide fascinating glimpses into the life of this remarkable man, his bright and incisive mind, his amazing curiosity, his influence on other people and his love of travel.

Part II contains seven papers on the methodology of modeling and simulation. Included are a reprint of a classical paper by Karplus himself, dealing with the range of models from systems that are well understood to those whose mathematical models are either imprecise or nonexistent. There are also reprints of papers by his colleagues Bekey and Korn, a paper on inverse problems by former student Rao Vemuri, a paper on discrete event representation of continuous models, an area of great interest to Prof. Karplus, a paper on characterization of ill-defined systems, and a survey of simulation languages.

Part III deals with applications of simulation, ranging from environmental systems to the study of calcium dynamics in heart muscle and the use of simulation and virtual reality to study brain aneurysms. Also included in this part is a paper on collaborative visualization by another former student. A paper on neural networks and their medical applications is included, since Dr. Karplus was influential in the creation of the IEEE Neural Networks professional society. Finally, the closing chapter deals with intelligent human-computer interfaces. The areas of visualization, virtual reality, and intelligent interfaces were Prof. Karplus's research emphases at the end of his life.

The editors would like to acknowledge the financial support of the UCLA School of Engineering and Applied Science and the UCLA Computer Science Department. We are very grateful for the assistance of Mr. Arun Bhadoria, who formatted the entire manuscript and prepared the Index. Without his help, this book might not exist.

Walter Karplus's death was indeed a major loss to the profession. We hope that this sampling of research areas influenced by his work will serve as a lasting memory of his contributions to modeling and simulation.

George A. Bekey and Boris Y. Kogan
Los Angeles

PART I

W.J. Karplus,

Scientist, Educator and Friend

Doctoral Students of Professor Walter J. Karplus

Name	Dissertation Title	Awarded
A. Odeh	Effect of Viscosity Ratio on Relative Permeability	Jan 1959
G. Rahe	Adaptive Sampling	Aug 1965
R.C. Wood	Optimum Quantizing in Hybrid Computations	Aug 1966
J.A. Howard	Partial Flux Switching in Multi-Aperture Ferrite Devices	Aug 1967
V.R. Vemuri	Identification of Nonlinear Parameters in Distributed Systems: A Hybrid Computer Approach	Jun 1968
A.F. Cardenas	A Problem Oriented Language and A Translator for Partial Differential Equations	Nov 1968
W. Cox	Interactive Computer Graphics for Engineering Field Problems	Jun 1972
O.P. Singla	Analog Subroutines for the Finite Element Method	Jun 1972
W. Tam	Digital Simulation Languages for Distributed Systems Parameter Identification	Jun 1972
T. Simundich	System Characterization: A Pattern Recognition Approach	Dec 1975
C.M. Lam	A Digital Simulation Language for Environmental Systems	Sep 1976
K. Apenyo	Segmentation of the Domain of Distributed Parameter Systems	Sep 1979
M. Milivojcevic	Physical Basis for Finite Element Methods	Sep 1979

Name	Dissertation Title	Awarded
D. L. Hudson	Rule-Based computerization of Emergency Room Procedures Based on Criteria Mapping	Jun 1981
Y. Shibata	Pattern Recognition Methods for Identification of River Pollution	Dec 1984
S. Cheung	Distributed Computer Simulation of Data Communications Networks	Dec 1986
S. Dimitriadis	A Methodology for the Multiprocessor Solution of Highly Oscillating Lightly Damped ODE Systems With Special Interest in the Transient	Dec 1986
A. Makoui	Data Flow Method for Simulation of Dynamic Systems	Dec 1986
H.C. Yeh	Application of a Parallel Processing Modular Simulation Systems for Dynamic Simulation of Nuclear Power Plant	Jun 1987
J. Betser	Performance Evaluation and Prediction for Large Heterogeneous Distributed Systems	Dec 1988
A. Pang	On Simulating and Visualizing Nonlinear Distributed Parameter Systems Using Massively Parallel Processors	Sep 1990
X. Lin	Boltzmann: An Object-Oriented Particle Simulation Programming System	Dec 1990
H.S.N. Dai	System Identification Using Neural Networks	Mar 1992
M. Stiber	Dynamics of Synaptic Integration	Jun 1992

Name	Dissertation Title	Awarded
W. Forniciari	Real-Time Simulation of Continuous System Using Heterogeneous Multiprocessors	Dec 1993
A. Wieland	Modeling, Simulation, and Control of Two-Legged Walking	Jun 1994
M. Harreld	Brain Aneurysm Blood Flow: Modeling, Simulation, VR Visualization	Jun 1996
M. Karpoukhin	The Analysis and Synthesis of Simplified Mathematical Models of Cardiac Action Potential for Wave Propagation Study on Massively Parallel Computer	Jun 1996
M. Orosz	A Methodology for Developing and Controlling Human/Computer Interface	Apr 1999
D.A. Lee	Path Line Decomposition for Visualization 3D Unsteady Vector Fields	May 2001
D. Liu	Data Management for virtual Reality Visualization of Large Flow Data Sets	Sep 2001
S. Golish	A Fast Nonlinear Method for Parametric Imaging by Positron Emission Tomography	Mar 2002

Publications of Professor Walter J. Karplus

A. Books, Chapters in Books and Editorships

[1] Karplus, W.J., *Analog Simulation: Solution of Field Problems,* McGraw-Hill Book Company, New York, 434, 1958. Russian Translation, 1962.

[2] Karplus, W.J., Editor of Analog Computer Section and author of 3 chapters in *Handbook of Automation Computation and Control,* Grable, E.M., Ramo, S., and Wooldridge, D.E., eds., Wiley and Sons, Inc., New York, Vol. 2, 1959.

[3] Karplus, W.J., Soroka, W., *Analog Methods: Computations and Simulation,* McGraw-Hill Book Company, NY, 483, 1959.

[4] Karplus, W.J., "Analog and Digital Techniques Combined", in *Computer Control systems Technology,* Leondes, C.T., ed., McGraw-Hill Book Company, New York, 141-167, 1961.

[5] Karplus, W.J., "Special Methods for the Solution of Field Problems" in *Computer Handbook,* Huskey, H.D., Korn, G.A., eds., McGraw-Hill Book Company, New York, Section 9, 69-85, 1962.

[6] Tomovic, R., Karplus, W.J., *High Speed Analog Computers,* Wiley & Sons, Inc., New York, 1962; Russian Translation 1964; Dover Publications, Inc., 252, 1971.

[7] Karplus, W.J., "Analog Elements", In *System Engineering Handbook,* Macol, R.E., ed., McGraw-Hill Book Company, New York, 10.1-10.19, 1965. (Handbook)

[8] Karplus, W.J., Ed., *On-Line Computing,* McGraw-Hill Book Company, New York, 336, 1966; Russian translation 1968; Romanian translation 1970.

[9] Karplus, W.J., Vemuri, V., "Numerical Solution of Partial Differential Equations", in *Digital Computer User's Handbook,* Klerer, M., and Korn, G.A., eds., McGraw-Hill Book Company, New York, 2.163-202, 1967.

[10] Girerd, J., Karplus, W.J., *Traitment des Equations Differentielles sur Calculateurs Electroniques,* in French, Gauthier-Villar, Paris, 559, 1968.

[11] Bekey, G., Karplus, W.J., *Hybrid Computation,* Wiley & Sons, Inc., NY, 464, 1968; Russian translation 1970; German Translation 1970; Japanese translation 1971; Polish translation 1976.

[12] Karplus, W.J., "On-Line Computing and Graphic Display", in *Computer Science*, Cardenas, A.F., Presser, L., and Marin, M.A., eds., Wiley and Sons Inc., New York, 223-240, 1972.

[13] Karplus, W.J., "Analog Signals and Analog Data Processing", *Encyclopedia of Computer Science and Technology*, Marcel Dekker, Inc., New York 2:22-51, 1975.

[14] Tomovic, R., Karplus, W.J., *Ogranicenja Formalne Teorije Upravljanja Sistema,* (Limitations of the Formal Theory of Large Systems), Gradeviska Knijga, Belgrade Yugoslavia, in Serbian, 86, 1979.

[15] Vemuri, V., Karplus, W.J., *Digital Computer Treatment of Partial Differential Equations,* Prentice Hall Inc., Englewood Cliffs, NJ, 449, August 1981.

[16] Karplus, W.J., Ed., *Peripheral Array Processors,* Society for Computer Simulation, La Jolla, CA, 11(1): 170, 1982; and 14(2):157, 1984.

[17] Karplus, W.J., Ed., *Multiprocessors and Array Processors*, Society for Computer Simulation, San Diego, CA 18(2): 321, 1987.

[18] Karplus, W.J., "Vector Processors And Multiprocessors", *Parallel Processing For Supercomputers And Artificial Intelligence*, K. Hwang, D. Degroot, Editors, Chapter 1, McGraw Hill Book Company, New York, 3-30, 1989.

[19] Karplus, W.J., Ed., *Multiprocessors And Array Processors*, Society For Computer Simulation, San Diego, California, Simulation Series, 21(1): 188, 1989.

[20] Skrzypek, J., Karplus, W., Eds., *Neural Networks In Vision And Pattern Recognition,* World Scientific Publishing Company, New Jersey, 209, 1992.

[21] Karplus, W.J., *The Heavens Are Falling: The Scientific Prediction Of Catastrophes In Our Time,* Plenum Publishing Company, New York, 320, 1992.

B. Papers Published in Professional and Scholarly Journals

[1] Karplus, W.J., "The Use of Electronic Analog Computers with Resistance Networks Analogs," *British J. of Applied Physics,* 6:356-357, 1955.

[2] Karplus, W.J., Allder, J.R., "Atmospheric Turbulent Diffusion from Infinite Line Sources: An Electric Analog Solution," *J. of Meteorology,* 13:583-586, 1956.

[3] Karplus, W.J., Smith, O.J.M., "The Application of Electrical Transients to Well Logging," *Transactions AIME,* 207:53-56, 1956.

[4] Karplus, W.J., "Water Coning before Breakthrough - an Electronic Analog Treatment," *AIME Transactions,* 207:240-245, 1956.

[5] Karplus, W.J., "Synthesis of Non-PR Driving Point Impedance Functions Using Analog Computer Units," *Transactions of IRE,* Professional Group on Circuit Theory, CT-4:170-172, 1957.

[6] Karplus, W.J., "An Electric Circuit Theory Approach to Finite Difference Stability," *Transactions AIEE, Communication and Electronics*, 210-213, May 1958.

[7] Karplus, W.J., "A New Finite Difference Stability Criterion," *Electrical Engineering*, 77:622, July 1958.

[8] Karplus, W.J., Bekey, G., Pekrul, P.J., "Atmospheric Diffusion of Air Pollutants," *J. of Industrial and Engineering Chemistry*, 50:1657-1661, November 1958.

[9] Karplus, W.J., Stephens, P.A., "Application of Finite Integral Transforms to Analog Simulations," *Transactions AIEE, Communications and Electronics*, 152-155, 1959.

[10] Karplus, W.J., "Simulation of Field Problems," *Instruments and Automation*, 32:253-255, February 1959.

[11] Karplus, W.J., "The Discrete-Space-Discrete-Time Computer," *Automatic & Remote Control*, Butterworth, Inc., London, 1:137-140, 1962.

[12] Karplus, W.J., "Trends in Field Simulation," *Annals of the Intl. Assoc. for Analog Computation,* V:1-3, January 1963.

[13] Karplus, W.J., "Hybrid Computer Technique for the Solution of Engineering Problems," *Bulletin Intl. Computation Center*, Rome, Italy, 3:22-29, January 1964.

[14] Karplus, W.J., "Design and Error Analysis of High Accuracy DC Voltage Measuring System," *Transactions of the IEEE*, PGIM, IM: 139-145, September 1964.

[15] Karplus, W.J., "A Hybrid Computer Technique for Treating Nonlinear Partial Differential Equations," *IEEE Transactions*, PGEC EC-13:597-605, October 1964.

[16] Vidal, J., Karplus, W.J., Keludjian, G., "Sensitivity Coefficients for the Correction of Quantization Errors in Hybrid Computer Systems," *Sensitivity Methods in Control Theory*, Pergamon Press, Oxford, 1966, p. 197-208.

[17] Karplus, W.J., "Error Analysis of Hybrid Computer Systems," *Journal of the Society for Computer Simulation,* 6:120-136, February 1966.

[18] Stoike, D., Karplus, W.J., "Heat Transfer in Pyrolytic Graphite in a Re-entry Environment: A Hybrid Computer Simulation," *J. of Spacecraft and Rockets,* 5:1491-1493, December 1968.

[19] Vemuri, V., Karplus, W.J., "Identification of Nonlinear Parameters of Ground Water Basins via Hybrid Computation," *J. of Water Resources Research*, American Geophysical Union, 5:172-185, February 1969.

[20] Cardenas, A., Karplus, W.J., "PDEL - A Language for Partial Differential Equations", *Communications of ACM,* 13:134-191, March 1970.

[21] Karplus, W.J., Russell, R., "Increasing Digital Computer Efficiency with the Aid of Error-Correcting Analog Subroutines," *Transactions On Computers, IEEE,* C-20:831-837, 1971.

[22] Karplus, W.J., "The Four Ages of Technology--And Simulation," *Simulation Today,* 6:21-24, 1972.

[23] Tam, W.C., Karplus, W.J., "The Use of the Digital Simulation Language PDEL in Hydrological Studies," *American Water Resources Assoc., Water Resources Bulletin,* 9:1100-1111, December 1973.

[24] Tam, W.C., Karplus, W.J., "Parameter Identification in Models of Hydrologic Systems Using the Digital Simulation Language PDEL," *Journal of the Society for Computer Simulation*, 22: 33-38, February 1974.

[25] Karplus, W.J., "Peripheral Processors for High-Speed Simulation," *Journal of the Society for Computer Simulation,* 29(5):143-153, November 1977.

[26] Karplus, W.J., "The Certification of Computer Science Educational Software," SIGCSE Bulletin, *Assoc. for Computing Machinery,* 10(1):76-78, February 1978.

[27] Karplus, W.J., Cohen, D., "Architectural and Software Issues in the Design and Application of Peripheral Array Processors," *IEEE, Computer,* 14:11-17, September 1981.

[28] Apenyo, K.F., Karplus, W.J., "A Domain Segmentation Approach to Modeling Distributed Parameter Systems," *IEEE Transactions on Systems, Man and Cybernetics,* SMC-12: 299-307, May/June 1982.

[29] Wu, C., Barkan, B., Karplus, W.J., Caswell, D., "Synthetic Aperture Radar Data Reduction Using Parallel Programmable Array Processors," *IEEE Transactions on Geoscience and Remote Sensing,* GE-20: 352-358, July 1982.

[30] Karplus, W.J., "Selection Criteria and Performance Evaluation Methods for Peripheral Array Processors," *Journal of the Society for Computer Simulation,* 43(3): 125-131, September 1984.

[31] Shibata, Y., Karplus, W.J., "Distributed Source Identification in River Pollution Systems by Pattern Recognition," *Transactions of the Society for Computer Simulation,* 3(1): 1-29, January 1986.

[32] Karplus, W.J., Shibata, Y., "The Application of Small Peripheral Array Processors to the Modeling of Distributed Parameter Systems," *Journal of the Society of Computer Simulation,* 46(6): 231-238, June 1986.

[33] Karplus, W.J., "The Emerging Role of Multiprocessor Systems in Simulation," *Journal of the Japan Society for Simulation Technology,* Tokyo, Japan, 5(3): 13-20, July 1986.

[34] Makoui, A., Karplus, W.J., "ALI: A CSSL/Multiprocessor Software Interface," *Journal of the Society for Computer Simulation,* 49(2): 63-71, August 1987.

[35] Dimitriadis, S., Karplus, W.J., "Algorithms For Dynamic Simulations of Space Structures," *Transactions of the Society For Computer Simulation,* 5(1):1 -25, January 1988.

[36] Yeh, H.C., Kastenberg, W.E., Karplus, W.J., "Applications Of Parallel Processing Modular Simulation System for Dynamic

Simulation of Nuclear Power Plants," *Nuclear Technology,* 84(1): 23-32, January 1989.

[37] Yeh, H.C., Kastenberg, W.E., Karplus, W.J., "Dynamic Nuclear Power Plant Simulations Using Single-Board Peripheral Array Processors," *Simulation, Journal of the Society for Computer Simulation,* 53(5): 209-217, November 1989.

[38] Dimitriadis, S., Karplus, W.J., "Multiprocessor Implementation of Algorithms For Ordinary Differential Equations," *Simulation, Journal of the Society for Computer Simulation,* 55: 236-246, October 1990.

[39] Kogan, B.Y., Karplus, W.J., Pang, A.T., "Simulation Of Nonlinear Distributed Parameter Systems On The Connection Machine," *Simulation, Journal of the Society for Computer Simulation,* 55: 271-281, November 1990.

[40] Karagueuzian, H.S., Kogan, B.Y., Khan, S.S., Denton, T.A., Karplus, W.J., Mandel, W.J., Diamond, G.A., "Induction Of Cellular Chaos During Quinidine Toxicity," *Journal of Electrocardiology,* 24(Supplement): 91-96, 1991.

[41] Kogan, B.Y., Karplus, W.J., Billett, B.S., Pang, A.T., Karagueuzian, H.S., Khan, S.S., "The Simplified FitzHugh-Nagumo Model With Action Potential Duration Restitution: Effects On 2D Wave Propagation," *Physica D,* 50: 327-340, 1991.

[42] Betser, J., Avritzer, A., Carlyle, J.W., Karplus, W.J., "A Cost/Performance Prediction Strategy For Shared-Load Heterogeneous Networks," *Transactions Of The Society For Computer Simulation,* 8(3):165-180, 1992.

[43] Kogan, B.Y., Karplus, W.J., Billett, B.S., Stevenson, W.G., "Excitation Wave Propagation Within Narrow Pathways: Geometric Configurations Facilitating Unidirectional Block and Reentry," *Physica D, 59*: 275-296, 1992.

[44] Tisdale, E.R., Karplus, W.J., "System Identification With Artificial Neural Networks," *International Journal of Pattern Recognition and Artificial Intelligence,* 6(1): 93-111, April 1992.

[45] Skrzypek, J., Karplus, W., Edited, Special Issue: "Neural Networks In Vision And Pattern Recognition," *International Journal Of Pattern Recognition And Artificial Intelligence*, 6(1): iii-viii, April 1992.

[46] Bekey, G.A., Liu, H., Tomovic, R., Karplus, W.J., "Knowledge-Based Control of Grasping in Robot Hands Using Heuristics From Human Motor Skills," *IEEE Transactions on Robotics and Automation,* 9(6):709-722, December 1993.

C. Papers Published in Proceedings or Records of Conferences & Symposia

[47] Karplus, W.J., "Electronic Analog Solution of Free Surface Problems," AIEE Technical Paper CP 5-149, *American Institute of Electrical Engineers*, New York, New York, 12, February 1956.

[48] Keansley, J.C., Karplus, W.J., "Study of Photochemical Reactions in the Atmosphere Related to Air Pollution Prediction," *Proc. of the Natl. Meeting of the Amer. Meterological Soc.*, Chicago, IL, February 1957.

[49] Karplus, W.J., "A New Active-Passive Network Simulator for Transient Field Problems," *Proc. of the IRE*, January 1961, 49:268-275. Russian Translation, 1961.

[50] Tomovic, R., Karplus, W.J., "Land Locomotion-Simulation and Control," *Proc. of the 3rd Intl. Congress of the Intl. Assoc. for Analog Computations*, Opatija, Yugoslavia, September 1961, 385-390.

[51] Karplus, W.J., Kovach, L., "The Use of Computers in Analysis". *Proc., Spring Joint Computer Conf.*, San Francisco, May 1962, 225-234.

[52] Karplus, W.J., "Analog Subroutines for Digital Computer Programs," *Proc., Intl. Symposium on Data Processing Machines*, Academy of Sciences, Prague, Czechoslovakia, September 1964, 178-190. (Reprinted in Simulation, 4:145-150, March 1965).

[53] Karplus, W.J., Rieman, F., Kanus, K., "Hybrid Solution of Transient Elastic Beam Problems," *Proc. of the 4th Intl. Congress, Intl. Assoc. for Analog Computation*, Brighton, England, September 1964, 516-520.

[54] Karplus, W.J., Howard, J.A., "A Transfluxor Analog Memory Using Frequency Modulation," *Proc. of the Fall Joint Computer Conf.*, San Francisco, 26:673-683, November 1964.

[55] Karplus, W.J., Vidal, J., "Characterization and Evaluation of Hybrid (Analog-Digital) Systems," *Proc. of IFAC Symposium*, Tokyo, 211-217, 1965.

[56] Karplus, W.J., Vidal, J., "Characterization and Compensation of Quantization Errors in Hybrid Computers Systems by Means of Sensitivity Analysis," *IEEE Intl. Convention Record*, New York, III: 236-241, March 1965.

[57] Karplus, W.J., Kanus, K., "Convergence Considerations in Hybrid Computation," *Proc. IFIP Congress*, NY, 2:403-404, May 1965.

[58] Karplus, W.J., Wood, R., "Error Analysis of Digital Computer Oriented Hybrid Systems," *Proc., IFIP Congress*, New York, 2:428-430, May 1965.

[59] Tomovic, R., Karplus, W.J., Vidal, J., "Sensitivity of Discrete-Continuous Systems," *Proc. 3rd IFAC Congress*, London, England, 40E.1-4OE.5, June 1966.

[60] Karplus, W.J., Vemuri, V., "Heuristic Optimization and Identification in Hybrid Field Simulations," *Proc. of the 5th Intl. Congress*, Intl. Assoc. for Analog Computation, Lausanne, Switzerland, 345-350, August 1967.

[61] Vemuri, V., Karplus, W.J., "Hybrid Computer Simulation of Ground Water Basins," *Proc. of the Symposium on Ground Water Hydrology*, American Water Resources Assoc., San Francisco, CA, 289-300, November 1967.

[62] Karplus, W.J., Bekey, G., "The Changing Role of Analog and Hybrid Computer Systems," *Proc. of the 1968 Intl. Fed. of Information Processing Congress*, Edinburgh, Scotland, 174-183, August 1968.

[63] Karplus, W.J., "Error Analysis in Hybrid Field Simulation," *Proc. AFIPS, Fall Joint Computer Conf.*, San Francisco, 15-16, November 1968.

[64] Rahe, G., Karplus, W.J., "An Adaptive Sampling System for Hybrid Computation," *Proc., Fall Joint Computer Conf.*, San Francisco 1225-1232, December 1968.

[65] Hara, H., Karplus, W.J., "Application of Functional Optimization Techniques for the Serial Hybrid Computer Solution of Partial Differential Equations," *Proc. of the Fall Joint Computer Conf.*, San Francisco, 565-574, December 1968.

[66] Cardenas, A., Karplus, W.J., "Design and Organization of a Translator for a Partial Differential Equation Language," *Proc. of Spring Joint Computer Conf.*, Atlantic City, NJ, 36:513-524, May 1970.

[67] Shih, J., Karplus, W.J., "Turbulent Diffusion in the Lower Atmosphere: A Hybrid Simulation Study," *Proc., Intl. Assoc. for Analog Computation*, 12:130-137, July 1970.

[68] Karplus, W.J., Cardenas, A., "Operating Experience with the Partial Differential Equation language PDEL: An Alternative to Hybrid Computation," *Proc. of the 6th Intl. Congress*, Intl. Assoc. for Analog Computation, Munich, Germany, 724-729, September 1970.

[69] Karplus, W.J., "Computational Tools for the Modeling of "Large" Distributed Parameter Systems," *Proc. AICA Symposium on Simulation of Complex Systems*, Tokyo, Japan, A-3/1 - A-3/7, September 1971.

[70] Karplus, W.J., Dracup, J.A., "The Application of the DSDT Hybrid Computer Method to Water Resources Problems," *Proc., 8th Annual Water Resources Conf.*, Rolla, MO, October 1972.

[71] Karplus, W.J., "System Identification and Simulation - A Pattern Recognition Approach," *Proc., Fall Joint Computer Conf.*, Anaheim, CA, December 1972.

[72] Karplus, W.J., "New Simulation Approaches to the Modeling of Environmental Systems," *Proc., Summer Computer Simulation Conf.*, Montreal, Canada, 57-59, July 1973.

[73] Nilson, R.N., Karplus, W.J., "Continuous System Simulation Languages: A State of the Art Survey," *Proc., 7th Intl. Congress*, Intl. Assoc. for Analog Computation, Prague, Czechoslovakia, August, 1973; reprinted in Annals of the Intl. Assoc. for Analog Computation, 16:17-25, January 1974.

[74] Karplus, W.J., "Analog and Hybrid Computing Modules for Signal Processing," *IEEE, NEREM, 1973 Convention Record* Boston, MA, 15(2): 79-88, November 1973.

[75] Tam, W.C., Karplus, W.J., "PDEL-ID: An Extension of PDEL for Distributed Parameter System Identification," ACM, *Proc. of the Symposium on Very High Level Languages*, Santa Monica, CA, 82-90, March, 1974.

[76] Karplus, W.J., "The Impact of Recently Developed Hybrid Computing Modules on Real-Time Signal Processing," AGARD, *Proc. of the 27th Technical Meeting of the Avionics Panel on "Real-Time Computer Based Systems"*, Athens, Greece, 26/1-26/10, May 1974.

[77] Karplus, W.J., "Hybrid Methods for Partial Differential Equations," *Proc., Summer Computer Simulation Conf.*, Houston, TX, 945-946, July 1974.

[78] Karplus, W.J., "The Role of Continuing Education in Computer Science," *Proc., 1975 Natl. Computer Conf.*, Anaheim, CA, 44:897-899, May 1975.

[79] Simundich, T.M., Karplus, W.J., "Feature Extraction Methods for the Modeling of Distributed Parameter Systems," *Proc., Summer Computer Simulation Conf.*, San Francisco, 98-103, July 1975.

[80] Karplus, W.J., "The Place of Systems Ecology Models in the Spectrum of Mathematical Models," *Proc., Soc. for Computer Simulation*, 5:20-23, December 1975.

[81] Karplus, W.J., "The Future of Mathematical Models of Water Resources Systems," *Proc., IFIP, Conf. on Biosystems Simulation in Water Resources and Waste Problems*, Bruges, Belgium, G. Vansteenkiste, ed., Systems Simulation in Water Resources, North-Holland Publishing Co., 11-18, 1976.

[82] Karplus, W.J., "Mathematical Modeling of Environmental Systems," *Proc. of the IFIP Conf. on Environmental Systems*, Tokyo, 3-8, April 1976.

[83] Karplus, W.J., "The Spectrum of Mathematical Modeling and Systems Simulation," *Proc. of the 8th Intl. Congress*, IMACS, Delft, Netherlands, 5-14, August 1976. Reprinted in *Mathematics and Computers in Simulation, Transactions of IMACS*, 19-1:3-10, March 1977.

[84] Karplus, W.J., "The Impact of New Computer Systems Architecture on the Simulation of Environmental Systems," *Proc. of the Conf. on Modeling and Simulation of Land, Air, and Water Resources Systems*, Intl. Fed. Of Information Processing, Ghent, Belgium, August 1977, North-Holland Pub. Co., 1001-1009, 1978.

[85] Karplus, W.J., "New Computer Structures for Low-Cost High-Speed Simulation," *Proc. of the 3rd Jerusalem Conf. on Information Technology*, Jerusalem, August 1978.

[86] Simundich, T.M., Karplus, W.J., "Applications of Pattern Recognition Methods for Modeling of Distributed Parameter Systems," (in Russian), *Automation and Computing Methods*, Minsk, 9:11-18, 1979.

[87] Karplus, W.J., "Simulation Using Array Processors: Software Issues," *Proc. - Simulation '80, Int. Assoc. of Science and Technology for Development*, Interlaken, Switzerland, June 1980.

[88] Karplus, W.J., "Software for Distributed System Simulation," *Proc. - Simulation '80, Int. Assoc. of Science and Technology for Development*, Interlaken, Switzerland, June 1980.

[89] Barkan, B., Wu, C., Karplus, W.J., "Application of Parallel Array Processors for SEASAT SAR Processing," *Proc. of the IEEE 1981 Int. Geoscience and Remote Sensing Symposium, (IGARSS '81)*, Washington, D.C., 541-547, June 1981.

[90] Karplus, W.J., "The Role of Pattern Recognition Methods in the Mathematical Modeling of Distributed Parameter Systems," *Proc. of the IASTED Symposium on Simulation, Modeling and Development*, Cairo, Egypt, September 1981.

[91] Karplus, W.J., Makoui, A., "The Role of Data Flow Methods in Continuous Systems Simulation," *Proc. of the Summer Computer Simulation Conf., Society for Computer Simulation*, Denver, CO., 13-16, July 1982.

[92] Karplus, W.J., "Parallelism and Pipelining in High-Speed Digital Simulators," *Proc. of the 10th IMACS World Congress*, 1:272-274, Montreal, Canada, August 1982.

[93] Karplus, W.J., "Peripheral Array Processors: Selection and Evaluation," *Proc. of array Processor Conference*, Society for Computer Simulation, San Diego, CA. 1-12, October 1982.

[94] Karplus, W.J., "The Spectrum of Mathematical Models," *IBM Perspectives in computing* 3:4-13, May 1983.

[95] Makoui, A., Karplus, W.J., "Data Flow Methods for Dynamic System Simulation: A CSSL-IV Microcomputer Network Interface," *Proc. of the 1983 Summer Computer Simulation Conference*, 376-382, Vancouver, Canada, July 1983.

[96] Ercegovac, M.D., Karplus, W.J., "On a Dataflow Approach in High-Speed Simulation of Continuous Systems," *Proc. of the Int'l.*

Workshop on High Level Computer Architecture, Los Angeles, CA, 2.1-2.8, May 1984.

[97] Karplus, W.J., Cheung, S., "Performance Evaluation Tools for Simulators Consisting of Networks of Microcomputers," *Proc. of the Summer Computer Simulation Conf. Society for Computer Simulation*, Vol. I, Boston, MA, 317-325, July 1984.

[98] Karplus, W.J., "The Changing Role of Peripheral Array Processors," *Proc. of the Conf. on Peripheral Array Processors*, Boston, MA, 1-13, October 11-12, 1984.

[99] Goodwin, W.R., Karplus, W.J., "An Innovative, Two-Week Computer Science Program For Employed Professionals," *Sixteenth SIGCSE Technical Symposium On Computer Science Education*, New Orleans, Louisiana, 17(1): 312-317, March 14-15, 1985.

[100] Cheung, S., Carlyle, J.W., Karplus, W.J., "Asynchronous Distributed Simulation of a Communication Network," *Proc. of the Summer Computer Simulation Conf.*, Reno, Nevada, 147-152, July 1986.

[101] Shibata, Y., Karplus, W.J., "A Pattern Recognition Approach to the Identification of Sources in Distributed Parameter Systems," *Proc. of the JSST Conference On Recent Advances In Simulation Of Complex Systems*, Tokyo, Japan, 93-98, July 15-17, 1986.

[102] Yeh, H.C., Kastenberg, W.E., Karplus, W.J., "Multi-Rate Integration Methods and Table Lookup Techniques Used in High Speed Dynamic Simulation of Nuclear Power Plants," *Proc. of the Second Int'l. Conf.on Simulation Methods in Nuclear Engineering*, Montreal, Canada, 768-788, October 1986.

[103] Karplus, W.J., "Parallelism and Pipelining: The Road to More Cost-Effective Scientific Computing," *Proc. of the Third Conference on Multiprocessors and Array Processors*, Volume 18, No. 2, San Diego, CA, 1-13, January 14-16, 1987.

[104] Tomovic, R., Bekey, G.A., Karplus, W.J., "A Strategy for Grasp Synthesis with Multifingered Robot Hands," *Proc. of the IEEE Int'l. Conf. on Robotics and Automation*, Raleigh, North Carolina, 83-89, March 1987.

[105] Dimitriadis, S., Karplus, W.J., "Automatic Generation of Ordinary Differential Equations Application Software For Multiprocessor Computers," *Proc. of the Multiprocessor and Array Processor Conf. (MAPCON IV)*, San Diego, CA, 52-57, January 1988.

[106] Lin, X., Karplus, W.J., "Overlapping Communications With Computations In Static Load-Balancing," *Proc. of the Multiprocessor and Array Processor Conference (MAPCON IV)*, San Diego, CA, 111-116, January 1988.

[107] Dimitriadis, S., Karplus, W.J., "Scheduling the Solution of Ordinary Differential Equations on Multiprocessor Computers," *Proc. of the Multiprocessor and Array Processor Conference (MAPCON IV)*, San Diego, CA, 58-66, January 1988.

[108] Lin, X., Chan, T.F., Karplus, W.J., "The Fast Hartley Transform On The Hypercube Multiprocessors," *Proc. of the Third Conference on Hypercube Concurrent Computers and Applications*, ACM Press, Pasadena, CA, 2:1451-1454, January 1988.

[109] Betser, J., Avritzer, A., Carlyle, J.W., Karplus, W.J., "Configuration Synthesis for A Heterogenous Backbone Cluster And a PC-Interface Network," *Proc. of IEEE Infocom'89*, Ottawa, Ont., Canada, 2:400-407, April 1989.

[110] Karplus, W.J., "Massively-Parallel Multiprocessing For System Modeling And Simulation," *Proc. of the International Conference On System Simulation And Scientific Computing*, Beijing, China, Pergamon Press, 1: 5-12, October 23-26, 1989.

[111] Lin, X., Karplus, W.J., "Boltzmann: An Object-Oriented Particle Simulation Programming System," *Proc. of the Second International Conference On Expert Systems For Numerical Computing*, West Lafayette, IN, 1-13, 1990. (Also reprinted in Intelligent Scientific Software Systems, J.R. Rice et al ed.)

[112] Pang, A.T., Karplus, W.J., Kogan, B.Y., "Simulation of Excitable Media on the Connection Machine," *Proc. of the SCS Multiconference on Distributed Simulation*, San Diego, CA, Simulation Series 22(2):77-82, January 1990.

[113] Kogan, B.Y., Karplus, W.J., Karaguezian, H.S., Khan, S.S., "Simulation of Reentry Processes In A Heart Muscle Using A Massively Parallel Computer," *Computers In Biomedicine: Proc. of the First International Conference*, Southampton, UK, 325-336, September 1991.

[114] Lin, X.A., Karplus, W.J., "Doing Physics Simulation In the Boltzmann Programming System," *Proc. of the SCS Multiconference On Object-Oriented Simulation*, Anaheim, CA, Society for Computer Simulation, 23(3): 110-117, January 1991.

[115] Kogan, B.Y., Karplus, W.J., Billett, B.S., Pang, A.T., Khan, S.S., Mandel, W.J., Karagueuzian, H.S., "The Role Of Diastolic Outward Current Deactivation Kinetics On The Induction Of Spiral Waves," *Proc. of the Ninth World Symposium On Cardiac Pacing And Electrophysiology*, Washington, D.C., PACE 14(4): Part II, 1688-1693, April 1991.

[116] Gluck, R., Dai, H.S., Karplus, W.J., "On-Orbit Nonlinear Structural Parameters Realization Via Artificial Neural Network," *Proc. of the American Institute Of Aeronautics And Astronautics*, 33rd Structures, Structural Dynamics and Materials Conference, Dallas, Texas, 2358-2366, 1992.

1

The Technical and Scientific Contributions of Professor Walter J. Karplus*

George A. Bekey
Department of Computer Science,
University of Southern California
Los Angeles, CA 90089
bekey@usc.edu

I would like to summarize the scientific and technical accomplishments of our friend and college worker Karplus. This is not an easy task, since he was a versatile researcher and prolific writer. Let me also apologize in advance to those of you who are neither engineers nor scientists for using some technical jargon; it is unavoidable in describing Walter's work, but I will try to provide some minimal explanations where possible.

I have had the privilege of collaborating with Walter since the late 1950s, for over 40 years. He was on my Ph.D. committee, we coauthored a book and a number of papers, the earliest in 1958 and the latest in 1993, and we met for breakfast about every two months for the past 20 years or more. Both of us are central European refugees; he from Austria and I from Czechoslovakia, and both of us have had a need to succeed in our new homeland; to work hard, and indeed to excel if possible. Dr. Karplus did this very well indeed. He authored or co-authored 9 books and some 130 technical papers in various areas. He directed the work of 31 Ph.D. students and countless master's students. He also edited a number of books and contributed to multiple scientific symposia. He gave numerous lectures on his research, both in the US and abroad. He taught at UCLA since 1955 and conducted numerous intensive short courses at various universities, including an annual course in simulation in Switzerland for perhaps 20 years. He traveled to Switzerland to teach this course this year, after being diagnosed with cancer. He was also highly sought as an expert witness in legal cases involving computer hardware and software; even in this area we collaborated and referred cases to each other, but I have to say he was much better at it than I. He received a number of honors, including Fulbright and Gugenheim

* This chapter is a slightly edited version of a talk given at a memorial seminar for Prof. Karplus at UCLA on November 25, 2001.

fellowships, awards from IFIP and NASA, and the title of Fellow in two professional societies.

Walter's early contributions were primarily in 4 areas: simulation, solution of field problems, hybrid computation, and array processors. More recently he has worked on human-computer interfaces, virtual reality (primarily as applied to medical problems) and neural networks. Let me try to summarize his work in these areas briefly.

Let me begin with simulation, since this area was the foundation of a great deal of his work. Simulation, as commonly used, refers to imitation and the use of models. For example, vinyl flooring may be simulated wood, there are simulated hamburgers made of soy, television and movies present simulations of life at earlier times. In engineering we use the word *simulation* to indicate a representation of some physical process in a form that's easy to study; this is also a model. Thus, we may simulate heat flow by electrical current. Or, the simulation may consist of solving the mathematical equations of the system on a computer. Walter has worked in both aspects of simulation, and he was respected worldwide as an authority on simulation.

Field problems concern such physical phenomena such as diffusion of heat in a material or propagation of communication signals in the atmosphere. Mathematically, they are described by partial differential equations that are difficult to solve on computers. This was particularly true in the early days of the computer field. In the 1950's and early 60's digital computers were programmed with punched cards and many mathematical models of physical systems were solved on analog computers. Walter published his first book in this field in 1958, with the title "Analog simulation: solution of field problems". This was a pioneering book. We published our first joint paper in this area in 1958, concerned with analog simulation of the diffusion of pollutants from a long line of automobiles. I still recall the newspaper article, that claimed "UCLA professors use computer to solve smog problem".

He continued to make contributions to the solution of field problems for many years. In 1981 he co-authored a book with his former student Rao Vemuri entitled "Digital computer treatment of partial differential equations". Note that whereas the 1958 book was concerned with analog approaches, by 1981 he was solving these equations on digital computers. Note also the word "treatment"...Walter liked that word and used it frequently in describing computer techniques. I used to joke about this with him, because I didn't think the equations were sick, so they did not need treatment. But then, words have many meanings, and Walter was very fluent in his use of English. He wrote a great deal, much of it by dictating to a tape recorder and having it transcribed. He wrote both papers and books this way. Of course, he did eventually switch to word processing on a desktop computer.

In addition to the book on solution of field problems by analog methods, in 1959 Walter co-authored a book on analog methods with another Walter, the late professor Walter Soroka from Berkeley and in 1962 one on high-speed repetitive analog computers with Rajko Tomovic, who at the time was a professor at the University of Belgrade in Yugoslavia. Tomovic was a frequent visitor to Walter's laboratory, having been here last year. He actually planned to visit UCLA again in April of this year, when he died suddenly in March.

The third area concerns hybrid analog-digital computation. Both Walter and I thought that combining the high speed of analog computers with the greater precision of digital systems, we could create outstanding new computers. I developed a hybrid system at TRW in the early 1960's. We taught summer short courses (Figure 1) in this field for a number of years, until it became clear that combining these two types of computers accentuated the drawbacks of both, rather than their advantages. Nevertheless, it was an exciting area and hybrid methods are now used in chip design. I mentioned Rajko Tomovic earlier. During the early 1960's Rajko collaborated extensively with Prof. Boris Kogan, a leading researcher in computers and control theory from the Institute of Control Sciences in Moscow, in the development of a hybrid computer. Prof. Kogan then came to UCLA to work with Walter in 1968. Later that year the Karplus-Bekey book "Hybrid Computation" was published (Figure 2). The following year, Boris Kogan translated the book into Russian. Walter visited Boris in Moscow in 1970, where he discovered to his pleasure that royalties from the book were deposited in an account in his name in Moscow, in rubles. While he could not remove the funds from the country, they gave him additional spending money. As many of you know, Prof. Kogan emigrated to the US in 1987 and has been on the UCLA faculty since.

The fourth area in which Walter made contributions deals with peripheral array processors. These were early attempts to create parallel computer systems. He edited several books for the Society for Computer Simulation in this field, between 1982 and 1989. While subsequent developments have made this approach obsolete, Walter was a pioneer and some of his approaches later found use in the analysis of performance of multiprocessor systems.

Figure 1. Announcement of Hybrid Computation short course, June 1963

Figure 2. Hybrid Computation book, 1968

More recently, Walter has worked on the design of intelligent interfaces between people and computers. This is an area of great importance, particularly in critical applications for NASA and the military, to allow operators to make decisions rapidly under stressful circumstances. Walter had support from NASA for this work and collaborated with cognitive psychologists for several years. A number of scientific papers arose from this collaboration, one of many interdisciplinary programs on which he worked.

Under support from NASA, he worked on applications of neural networks to control and identification of nonlinear systems.

He also worked on models of biological systems, including studies of the excitation of heart muscle, which he did jointly with Prof. Kogan and faculty in the Department of Cardiology. Another area involving cooperation among disciplines, concerns the application of virtual reality methods to the study of brain aneurisms. While I am not very familiar with this work, I know it involved the ability to visualize critical blood vessels in the brain where surgical techniques could be used to repair an aneurism, a point in which there may be a breakage of a the wall of an arteriole and subsequent leakage to a small vein. Walter collaborated with brain surgeons in the UCLA School of Medicine on these problems.

Because of his interests in virtual reality, intelligent interfaces and neural networks, he was elected President of the IEEE Neural Network Council, a highly prestigious professional society.

I hope these comments have given you a glimpse of the rich professional life of our friend and colleague and his numerous contributions to science and technology. He was a versatile computer scientist, with a great scientific curiosity coupled with a strong desire to assist people in need and society as a whole. In this latter area, you will hear from Dr. Vemuri about his work with the Eco Foundation. He was also on the board of the Wright Institute, an organization providing long-term psychotherapy for indigent persons. I have emphasized his scientific contributions, but I cannot leave out his teaching ability. He was a gifted and eloquent communicator. In fact, when both of us participated in short courses, he generally received higher evaluations from the students than I, and he deserved them.

Let me end with a comment on his latest book, "The Heavens are Falling". This is a popular book dealing with a variety of catastrophes predicted by various so-called experts, from total destruction of the atmosphere to overpopulation to nuclear annihilation. He points out in this book how people tend to extrapolate linearly from current experimental data to some distant future. This is also a problem with simulation models that depend entirely on experimental data rather than on first principles. He points out that when data collected under restricted conditions are used to make predictions under different conditions, one can obtain totally invalid or even ridiculous results. The book is optimistic in that it shows that it is likely the great catastrophes will not occur. On the other hand, he reminds us that most of us do not use models wisely and we will probably continue to make faulty predictions. This is Walter Karplus at his best: applying his knowledge of computers and science to contemporary social problems, with keen insight, humor and wisdom. Speaking of humor, Walter had a great reservoir of stories, but unfortunately I'm not permitted to tell any of them.

He has contributed so much and left us a great legacy. Those of us who had the privilege of working with him will always remember his remarkable understanding of physical problems, his ability to communicate and his wit. Walter, old buddy, we'll miss you, both in simulation and in reality. Thanks for what you have done for the profession and for us who were your friends and colleagues.

2

Dr. W. Karplus, My Friend and Colleague

Boris. Y. Kogan
Department of Computer Science,
University of California, Los Angeles,
CA 90095-1679 USA
kogan@cs.ucla.edu

It was a big privilege for me to have Walter as my good friend during last 41 years. For the first time, I met Walter personally was during the First International Congress on Automatic Control (organized by IFAC) in July 1960. Our meeting is commemorated in the Photograph (See Figure 1), taken together with some other members of the US delegation.

Figure 1

The congress was held in Moscow and hosted by Research Institute of Control Science, which then was recognized over the world as a leading center in the theory of feedback control systems.

At that time I was working in this Institute as a scientific director of the laboratory for computer simulation of the control systems and was very well familiar with Walter's first books: "Analog Solution of Field Problems" (1958) and with W. Soroka. "Analog methods: Computation and Simulation" (1959). The books were very popular in the former Soviet Union among researchers and PhD students due to clarity in the treatment of the difficult problems and originality of the presented results. The latter was the inherent characteristic of the whole of Walter's scientific and teaching activities.

In the following years we often met on different scientific occasions. In 1966 we met in Dubrovnik (in the former Yugoslavia) as teachers for the International School on Optimization Problems. The group of teachers can be seen in Figure 2: from left to right are Dr. Ljubo Radanovich (from Boris Kidrich Institute in the former Yugoslavia), Dr. Karplus, myself, Dr. Tomovich (Belgrade University, in the former Yugoslavia), Dr. Gamkrelidze (Institute of Mathematics, Academy of Sciences in the former Soviet Union).

Figure 2

In 1967, we met on International Congress of AICA (Association Internationale Pour le Calcul Amalogique) in Lausanne, Switzerland where Walter introduced me to his good friend and prominent scientist Dr. G. Korn (Arizona University). In 1968, I attended American National Computer Conference in Atlantic City and visited UCLA as Walter's personal guest (see Figure 3.). In 1970 Walter visited Moscow by invitation of the Soviet Union Academy of Sciences. During this visit I accompanied him on his trip to the Middle Asian republic of Uzbekistan where he not only met with local scientists but also saw a lot of interesting sights in the cities of Tashkent, Samarkand and Buchara.

Figure 3

At that time, the book "Hybrid computation "authored by George Bekey and Walter had already been translated into Russian. I was the scientific editor of the Russian translation of this innovative book. Later on that helped me significantly (as Principal Investigator) to accomplish the joint project of the new Hybrid Computer system HCS-100 with the Pupin Institute in Belgrade. Walter and I continued our scientific friendship by exchanging the papers and letters. I also had sent to his laboratory one of my best pupils and colleagues, the late Dr. A. Petrov. After 1973 when my brother immigrated to Israel I became persona non grata in the Institute where I worked for almost 40 years. Walter's letters were a great support for me in this difficult time. Finally, when I decided to emigrate Walter, G. Bekey and G. Korn were the people who helped me very much not only to arrive in Los Angeles, but also to enter to a new life here. Due to Walter's care I am working here in UCLA. For the past 14 years, he showed vivid interest to my research and participated in more than 10 publications. I attended several of Walter's lectures for students and his scientific presentations and was always greatly impressed. There were brilliant by performance, deep in content, and explained the complex subjects in a simple form. Walter was a great teacher, an outstanding scientist and a "Man" with a capital letter. I am proud that I had such a friend. He left a great legacy for all of us. We thank him for that, and will try to follow his example.

3

A Conversation with Walter J. Karplus

V. Rao Vemuri
Professor of Applied Science
University of California, Davis CA 94550
rvemuri@ucdavis.edu

I remember Prof. Karplus as a man of few words – very few words. In my forty some years of association with him, I recall only one occasion on which I had an extended conversation with him. I will come to that in a moment.

I first came to know about Walter in 1964. I still remember very vividly the occasion. I was attending evening classes at Purdue University Extension at Indianapolis while employed at RCA's Home Instrument's division. During one of my visits to the campus (now IUPUC), I happened by chance to see a book entitled High-Speed Analog Computation by R. Tomovic and W. J. Karplus in a glass display case. I jotted the name and address and wrote a letter to Dr. Karplus expressing my desire to work for a Ph. D. under his tutelage. Within a week, I received an application form with a note containing a "half-sentence scribble" urging me to apply.

Some time later, I received an admission letter. What do I do with it without some financial assistance? I pulled my courage, rehearsed what I was to say, prepared notes for any possible grilling and called Professor Karplus up. His secretary Maya answered the phone and told me that he was about to go to Europe and patched me through. Prof. Karplus picked up the phone and said, "Yes, there is money to support. You can come." With that he was about to hang up. I prevailed upon him to send me a written statement to that effect. This time, I lucked out; I received one full sentence! It simply stated, "You are appointed as a Research Assistant at $2.75/hour."

During my four years at UCLA, I can actually count the number of times I had a "conversation" with Prof. Karplus. They were infrequent and utterly brief. This is not to say that he was not interested in me. After passing my written qualifier examination, I wanted to go home to visit my parents in India; it had been four years since I last saw them. On the eve of my departure, he gave me R. S. Varga's Matrix Iterative Analysis and asked me to read it while in India. As I was leaving, he said, "Fill your time cards for the summer and leave them with Maya." That is vintage Karplus for you.

Now I would like to relate a really extended conversation I had with Walter. We were both attending a Neural Networks conference in Houston, Texas. I bumped into him and we had a cup of coffee in the hotel lounge. To my great surprise, Walter said, "Rao, let us go out for a brief walk so we can chat." I simply could not believe my ears. During that rambling conversation, he was talking about the purpose of life. We talked about J. Krishnamurti's philosophy and Indian (Hindu) approach to life. We talked about life and about death. He envied, he said, the Hindu way of life, "It prepares you so well for death."

He felt that he did not do enough; he could have done more. Was he referring to his contributions to the profession or to life itself? I was not sure. To cheer him up, I started reciting a litany of his accomplishments. I even took the liberty – for the first and only time – to point out how close he came to inventing what we now call "Hopfield net."

"Rao, let me tell you a story. This is a Kipling story. You probably know it already. I read a German translation of it when I was a kid.

"You see this was a story about an Indian bureaucrat. He rose to a high position in government, wielded a lot of power, felt he was "in charge" and eventually retired from service.

"In spite of a string of successes in his professional life, he was unhappy. Dissatisfied. He felt that there was a void in his life. He did not do enough. He could have accomplished more. That was the source of his anguish.

"He finally decided to go the Himalayas. He selected a secluded cave on a high mountain slope, sat cross-legged and began to meditate – in his search for satisfaction in life. In no time villagers in the valley below came to know of the new "swami." They visited him with reverence. Offered him flowers, fruit and milk. Sought his blessings. And they asked him for his words of wisdom.

"What could he offer? He himself didn't know what the question was. He nevertheless accepted their offerings and gave them a mystic smile.

"One day a big earthquake shook the whole valley floor. Mudslides and floods soon followed. The villagers ran helter-skelter. It would take weeks for any assistance to reach them from the government. The only thing they know how is to pray. And they did.

"The swami on the hilltop came to know about the plight of the villagers below. He instinctively knew what to do. He abandoned his seclusion and search for happiness. He ran down to the village. He organized the villagers into teams to dig through the rubble. He organized search and rescue missions. He found a purpose for his life. He was happy! A grateful village treated him like a God who descended from heaven."

Walter concluded his narration saying, "Rao, I cannot remember the ending. If you can ever find this story in its original please read it and let me know."

"Walter, do you know you can still find genuine swamis if you go to places like Rishikesh at the foothills of the Himalayas? I can come with you and act as your guide!"

"Rao, I dream about things like that. We westerners are too materialistic!"

4

Some Memories of Walter Karplus

John Mcleod
Founder Society for Modelling and Simulation,
8484 La Jolla Shores Drive, La Jolla, CA 92037
mcleod@sdsc.edu

Walter Karplus once observed "All technical writing should self-destruct after five years." On first reading that would seem a puzzling statement coming from one who himself was such an important contributor to the technical literature in his field. But knowing Walter, we understand now that he was only paying tribute to the speed with which advances in technology are being made, and not referring to archival-class material.

Early in 1992 when I was editing a column "Simulation in the Service of Society" for SIMULATION, technical journal of the Society for Computer Simulation, I was invited to write a review of a book, THE HEAVENS ARE FALLING: The Scientific Prediction of Catastrophes in Our Time, by Walter Karplus. It was a difficult request to fulfill because I found that I did not agree with some of my friend's points. But then, agreement with the author is certainly not a requirement for an unbiased review; in fact it might be detrimental to an honest appraisal of the work, so I took on the task.

In his book Walter gives his definition of a catastrophe: "We will use CATASTROPHE to mean an event, or series of events, having all of the following attributes: Very UNFAVORABLE for a substantial portion of human society; Extraordinarily INTENSE in magnitude; Exceedingly RARE in occurrence; VERY LONG-LASTING in its effects on SOCIETY.

Eight catastrophes, all predicted by eminent scientists, Walter explained, would be the focus of his attention:

1. Depletion of the ozone layer and the consequent increase of our exposure to ultra-violet radiation.

2. Change of the global climate due to increases in atmospheric carbon dioxide and other "greenhouse" gases.

3. Pollution of the atmosphere through the emission of industrial chemicals.

4. Pollution of the atmosphere by radioactive particles.

5. The AIDS epidemic.

6. Over-population and the population explosion.

7. Worldwide economic depression.

8. Earthquakes of catastrophic proportions.

In his book Walter explains that he has examined each of these catastrophes. This reviewer found the entire book intriguing in a "things I didn't know till now" sort of way, and the discussion of the eight catastrophes fascinating in a "murder mystery" sort of way, perhaps because, as the Dr. Karplus points out, death is a catastrophe for the victim, and no one knows how it will all end.

The organization of the book seems to reflect the Walter's engineering training. After an overall disclaimer, "CATASTROPHES CANNOT BE PREDICTED SCIENTIFICALLY," there is a similar pattern in the discussion of each of the eight catastrophes. A quotation is followed by a brief introductory discussion. Next there is a section on Theories, then Data.

After that the Causal Chain is presented. This was of particular interest to this reviewer because, although I must have been subliminally aware of the necessity for some cause-and-effect reasoning in both the development and validation of models predicting catastrophes, I never singled out the process that distinctly, nor pointed out its importance as clearly in my writings, as does Karplus. The importance of the Causal Chain is graphically emphasized by simple block-diagrams of each of the eight Catastrophes discussed.

In most cases Predictions are made, followed by a section What Must Be Done. Where important, the political aspects of the situation are discussed.

I found the Personal View of the erudite author, with which he closes his discussion of each Catastrophe, of great interest. As indicated by the many References and the Bibliography, he is well read on the subjects about which he writes--and thinks!

So what does he think? In the final chapter, titled What Does It all Mean? he tells us. However, it seemed to this reviewer that Dr. Karplus has made such a strong case for each of the catastrophes considered that the Big Picture which he presents in closing is relatively pallid.

"It is difficult to contend, in the light of the facts presented earlier" Walter says, "that we are not doomed to be finished off by at least one of the catastrophes described."

However, your reviewer grasps a straw of hope from something Walter previously passed over. Attached somewhat precariously, I thought, to his discussion of warming as the result of the Greenhouse Effect is a Postscript covering arguments predicting that "THE ICE AGE IS COMING"!

Now the predictions of both of these two catastrophes cannot be correct--one or perhaps both sets of arguments must be at least partially in error. Thus this optimistic reviewer can HOPE they all are, if not somewhat is error, at least--like the reports of Mark Twain's death--grossly exaggerated.

The foregoing comments were based on my review as published in SIMULATION, May 1992. Rereading the book in the light of what has actually occurred and current trends, I am impressed with the wisdom, and sound common sense, of certain statements. Having pointed out that we frequently hear "exhortations such as 'This is our last chance!' and 'We are at a turning point in history!'....In fact, the present moment is a turning point in history, as is every other moment but only because the wheel is going around and around all the time."

"We must accept the fact that a new major, imminent catastrophe will come out of the woodwork every few years--a predicted catastrophe that will have the full backing of a reputable community, that will get full attention from the media...that will call for a prompt reordering of priorities. And as our technology develops, as computers become more powerful, the breathing time between catastrophe predictions will become shorter and shorter. We must recognize that fact, and we must tune our response to catastrophe predictions to that fact."

Most of my contacts with Walter came at meetings and conferences of SCS (The Society for Computer Simulation) and AICA (Association Internationale pour le Calcul Analogique). I recall one occasion when Professor Henri Raymond, President of AICA, delivered his keynote address in French, with Walter as translator for the benefit of the American audience. Many of our contacts came about when Walter was President of SCS and active in the Publications Committee of the Society. However, I recall two instances that were purely social. Both took place far out of either of our usual environments, one in Tokyo and the other in Melbourne, Australia. In both cases we were attending conferences on computer simulation.

The first of these encounters was in a department store in Tokyo billed as the largest in the world. Suzette and I were just looking, but Walter had a clear objective. While in Tokyo he was staying with his Japanese wife's family. He had apparently come to sleeping terms with the Japanese futon, but not the hard blocks they use for pillows -- he was looking for a good soft U.S. style pillow.

Our next meeting was planned. After the conference sessions and a few drinks Walter, Tako, Suzette and I planned to go to dinner together but, not knowing Melbourne, we had no idea where. So, as most restaurants had menus posted where copies could be read from the street we decided to just walk and read menus. As we walked our hunger increased, not necessarily linearly. Midnight approached, but still no consensus as to what or where to eat. Tako must have had something in mind as she would walk on ahead and decide against places, sometimes even before we got there. I was stymied by dishes the names of which meant nothing to me. Suzette said she would go with the majority -- if there were one, while Walter was just being his agreeable self.

We finally settled for Mongolian barbecue. But I still don't know what we ate!

That, years ago, was the last time I saw Walter. But I still treasure a copy of an e-mail message to me which must be one of the last he ever wrote. In it he apologizes for being slow to answer my request for a writeup on how he first became involved in computers and simulation, on the grounds that he has had "some health problems." It was written just a few days before his death.

5

Walter Karplus – Friend

Ira Winn
Professor (Emeritus) Education and Urban Studies
California State University Northridge,
San Luis Obispo, CA 93408
iawinn@pacbell.net

Walter was a family friend, a hiking partner of more than thirty years, and a colleague, albeit from a different specialty and a different university. We had a kind of intellectual sparring relationship, testing our ideals and ideas as well as our values and biases, while laboring on trails from the Santa Monica and San Gabriel mountains to the high Sierras and his favorite haunts in Zermatt, Switzerland. Underlying our often humorous discussions, at home and in nature, was a foundation of mutual respect and comradeship.

Walter Karplus was a scientist with a strong liberal arts bent. He was very well read in fiction, biography, history, and travel. He could discourse about most any subject – football to game theory; geology to swimming, films to women's liberation to money and politics. What he knew in computer science I cannot even guess, but his stories of scientific adventure never ceased to astound me. Many years before personal computers became commonplace, he advised me to junk my electronic typewriter and buy an early Apple, an idea which sent me into gales of laughter. He must still be laughing last as he sees me peering into my monitor.

Politically and philosophically, we shared common insights and values. But we had some strong disagreements, one of which is worth relating because it shows both the strength of our friendship and his openness to change.

While he was writing THE HEAVENS ARE FALLING, a book about catastrophe theory and reality, aimed at a lay audience, we got into a huge discussion of global warming and the greenhouse effect. At that time he thought them a threat of relatively minor consequence. I had argued that glacial melting was far more serious than he thought, and that flooding of coastal cities could displace huge numbers of people. He responded that was unlikely and that people can always move to higher ground or even to Canada if need be. We abandoned the argument that day because neither of us would give an inch.

A few months later, as we met at a trailhead, he emerged from his car visibly agitated. It seems his brother had been caught in the Oakland Hills fire and he and his family had to flee with only the clothes they had on. The house was a total loss.

I feigned unconcern, and waved off his puzzlement with the comment, "Well, Walter, it's not a big deal. He can move to Canada and join the other refugees." "He gave me a sardonic look and we both burst out laughing. As I remember, he then did one of his famous "fadeouts", where he is physically with you but is mentally somewhere in space, mulling some puzzle or passing idea. You get used to that in the presence of genius (although my wife, Arlene, never did and invariably joked with Walter about his tendency to "abandon ship"—to be sometimes abrupt and distant.)

Then, when his book on catastrophe theory was press-ready, he asked me to look it over. I noticed that global warming or climate destabilization had moved way up the ladder of importance. But the final stroke was Walter's:

A journal editor had asked him to recommend an academic to review the book and he had no hesitation in putting forward my name, as I later learned. He knew we still had some differences in perspective. Yet, I did a critical but positive review for the JOURNAL OF CONTINGENCIES AND CRISIS MANAGEMENT (June 1993) and Walter, when he read it thanked me for the fairness of my appraisal. He was always a scholar and a gentleman.

It was very important to Walter that he stays *au courant* in almost every area of his interest. His travels were legendary. He had many causes, which he quietly supported. He loved to gamble and to beat whatever system confronted him. He was charitable and exceptionally giving to his wife and two children.

Sometimes he would be moody, depressed about the state of the world or his inability to control the flow of things –or bothered by the difficulty of connection to spiritual life, which both attracted and puzzled him. But he had an amazing ability to rebound and seize the moment. He was a master at rising to the occasion.

Walter approached and moved through death's door with awareness and with great dignity. Maintaining control meant everything.

I feel fortunate to have shared times with him. I'll always remember Walter Karplus not just as a friend, but as a seeker and a pioneer.

PART II

Methodology

6

The Spectrum of Mathematical Modeling and Systems Simulation*

Walter J. Karplus
*Computer Science Department, University of California,
Los Angeles, CA 90024*

Summary. The methodology involved in the modeling and simulation of physical, life and social science systems is viewed in perspective. A critical factor determining the validity of a model is the extent to which it can be derived from basic laws and insights into the internal structure of the system using deductive methods, rather than relying upon observations and measurements of the system input and outputs. Accordingly, the mathematical models as they arise in various application disciplines are arranged along a spectrum according to the relative amount of deduction and induction involved in their construction. This provides an insight into the ultimate validity of simulations and to what use they can properly be put.

1. Introduction

Recent years have seen intensive attempts to extend the art of mathematical modeling and systems simulation to an ever-expanding range of application areas. A particularly significant impetus toward this development of mathematical models has been the tendency of virtually all physical, life, and social science disciplines to become more quantitative in their methodology. The evolution of the modeling art has not been without difficulties and controversies. For example, specialists in the modeling of such "hard" systems as electromechanical control systems have challenged the validity of models in such "soft" areas as economics and sociology; similarly the relevance and utility of well-established mathematical techniques useful in modeling "hard" systems has been questioned by some of those engaged in modeling "soft" systems. In fact, doubts have been raised whether the term "model" really means at all the same thing in diverse areas of application. It is the purpose of this paper to help clarify some of the problems arising from

* Reprinted from *Mathematics and Computers in Simulation,* v. 19, Walter J. Karplus, The Spectrum of Mathematical Modeling and Systems Simulation, pp. 3-10, Copyright (1977), with permission from Elsevier Science. Research in Modeling at the University of California was supported by the National Science Foundation under Grant GK 42774.

the application of essentially similar concepts to fundamentally different real-life-systems, by providing a unified perspective of the modeling process and the variety of applications of mathematical models.

2. Assumptions Inherent in Modeling

The basic tenets of mathematical modeling have been so deeply inculcated upon its practitioners, that they are rarely if ever questioned in practice. It is useful, to review briefly the basic assumptions or paradigms which are the foundation of all mathematical modeling [1]. Models can be considered credible or valid only to the extent that the systems being modeled satisfy these successively more restrictive conditions. Of fundamental importance in all system studies is the notion of *separability.* To a greater or lesser extent, all objects and phenomena in the universe are interrelated. In defining a system for modeling and analysis, however, it is assumed that most of these interactions can be ignored, so that the system can studied as a separate entity. This separation frequently involves the definition of a "boundary" of the system or an enumeration of all of the elements or components comprising the system. Once a system has been defined as a separate entity, it is necessary to describe the interaction of the system with the external world. In order to arrive at a model of manageable proportions, it is necessary to invoke a *selectivity* condition. That is, it is necessary to assume that of all possible interactions, only a relatively small subset are relevant to a specific inquiry or purpose. For example, in modeling an electrical circuit, we ignore thermal, acoustic, optical, and mechanical interactions and consider only electrical variables. Thus the model of the system is expressed by the general relation

$$S \subset U \times Y \qquad ...(1)$$

where U and Y are sets of inputs and outputs. The selectively assumption implies that negligible errors are introduced in ignoring all other inputs.

Having invoked selectivity, it is now necessary in mathematical modeling to assume *causality.* From the set theoretic point of view, causality implies that the system inputs and outputs are related by a mapping function f: U→Y, so that

$$S : U \rightarrow Y, \qquad ...(2)$$

The subject of causality has been widely discussed by philosophers and all types of scientists. It continues to be a challenging and perplexing problem.

In general, causality can only be assumed if it is possible to identify a complete chain of causally related events linking the inputs and outputs. It is not enough to observe that Y invariably follows U; for Y and U may both be the results of a common cause [2–4]. In order to derive a model useful for quantitative prediction, it is necessary to introduce the concept of "state". This requires additional assumptions regarding the properties of the system. In particular it is necessary to assume that the inputs and outputs are measurable, i.e., capable of numerical representation. In addition, a number of restrictive assumptions relative to the system state are required in order to derive dynamic models. In mathematical modeling and in systems studies in general, there is a danger in becoming so immersed in the problems involved in determining and implementing the mapping function, f, that sight is lost of the fact that the validity of the model is subject to the applicability of all of the basic assumptions mentioned above. In the discussion below, it is assumed however that adequate consideration to these points has been given in formulating the modeling problems.

3. Modeling Methodology

The construction of a mathematical model of a system entails the utilization of two types of information: (1) knowledge and insight about the system being modeled; (2) experimental data constituting observations of system inputs and outputs. The utilization of the former class of information involves deduction, while modeling using empiric observations involves induction. A mathematical model constructed purely deductively can generally be considered a unique solution of the modeling problem. By contrast, when inductive methods are used there exist always an infinite number of models satisfying the observed input/output relationships. In inductive modeling it is therefore necessary to introduce additional assumptions or constraints to help select the optimum model from the infinitude of possible models. For this reason, in constructing a mathematical model an attempt is always made to carry deduction as far as possible, and to rely upon system observations and experiments only where absolutely necessary. Deduction is reasoning from known principles to deduce an unknown; it is reasoning from the general to the specific. In modeling deductively, one derives the mathematical model analytically by making use of a series of progressively more specific concepts which may be broadly categorized as: laws, structure and parameters. Laws are the basic principles which determine the general character of the equations characterizing the system. In physical systems, these laws are usually expressions of the principles of conservation and continuity— that matter and energy can only emanate from specified sources. The application of these laws permits the derivation of the familiar partial differential, ordinary differential and algebraic equations. The basic laws are usually formulated for specific application areas and include such well known principles as Kirchhoff's laws, Newton's laws, Fourier's laws, Maxwell's equations and the Navier—Stokes equation to mention just a few. The application of a law to a system usually involves the focusing of attention upon a single physical area. For example, when utilizing Kirchhoff's laws or Maxwell's equations in analyzing electrical systems, one ignores chemical and thermal processes which may be going on simultaneously within the system. The system being modeled is usually regarded as consisting of a large number of interconnected elements, components, or subsystems. This view of the system frequently involves the making of simplifying assumptions or approximations. However, because this approach lends itself so well to formal methods and eventual computer implementations, even systems which are continuous in space (as for example a water reservoir or the atmosphere) are often viewed as being comprised of an array of closely-spaced elements. The construction of a valid mathematical model demands the knowledge of the types of elements which are present in the system and how these elements are interconnected. The interconnections specify the paths over which matter or energy flows within the system; the types of elements determine what happens to this matter or energy at different locations within the system. Mathematically, this in turn determines the number of simultaneous equations in the mathematical model, as well as the types of terms in each equation, (first derivatives, second derivatives, integrals, etc.). Whereas deductive knowledge of the laws governing the system is obtained from the study of a scientific discipline, such knowledge about the structure of a specific system can only come from an insight into the specific system being modeled.

The parameters in a mathematical model are the numerical values assigned to the various coefficients appearing in the equations. These are related to the magnitudes of all the elements comprising the system as well as to the boundary and initial conditions, which together with the governing equations constitute completely-specified model. As in the case of structure, deductive knowledge of the parameters entails a priori knowledge or assumptions about the specific system under consideration. Where the laws, structure and parameters of a system are completely specified or known, the mathematical model characterizing the system can at least in theory be derived mathematically. To be sure the existence of nonlinearities, variable parameters l and changing structures can make this derivation exceedingly difficult, so that frequently numerical methods and approximations are called for. Induction, by contrast, is a much more uncertain process. Here a set of system inputs and outputs are observed, measured and recorded, and some or all of the mathematical model is to be inferred from them. Questions as to the quantity and quality of observations sufficient to this task therefore arise. System observations may be obtained either actively or passively. In active experiments, the modeler specifies interesting inputs, applies these to the system under study, and observes the outputs. By contrast, in passive observations one is unable to specify inputs but must accept whatever input/output data is available or can be gathered. In some modeling applications, for example electrical or mechanical systems, active experimentation is widely used and leads to relatively valid models. Such experimentation is a particularly powerful tool because key experiments can be constructed to answer specific questions about the system and to establish causal relationships between inputs and outputs. In many application areas such as for example environmental systems or economic systems, the prototype system is too large or too remote to permit the application of specified excitations.

4. Degrees of Inductivity

Mathematical modeling problems are often referred to as "black box" problems. In such problems the inputs and outputs are specified but the mathematical model of the system is unknown — hence it is considered a black box. In order to determine the contents of the black box, it is necessary to solve the inverse problem, imposing as constraints any available knowledge about the system. The more that is known about the system, in addition to observed inputs and outputs, the greater is the probability that the mathematical model will be a satisfactory representation of the prototype system — that it will be useful in predicting outputs other than those utilized in constructing the model. In most mathematical modeling situations, the contents of the "black box" are not totally unknown. In fact, occasionally the contents are completely known so that the entire mathematical model can be derived deductively. Sometimes the entire contents of the box are known except for a few element values or parameters; in other cases the assortments of elements within the box may be known, but not how they are interconnected, and in still other cases, there may be only a qualitative intuitive insight into the behavior of certain components of the system. It therefore appears appropriate to extend the "black box" metaphor so as to provide for boxes of various "shades of gray" — the lighter the shade of gray, the more deductive and the less inductive the mathematical modeling process. An examination of the models utilized in any specific application discipline reveals that the bulk of the models can be characterized as being of approximately the same shade of gray. In fact, the entire modeling methodology and the eventual utilization of the models is attuned to the uncertainty existing as to the mathematical representation of the system, S, at the outset of the modeling procedure. It is possible therefore to arrange the various fields which utilize mathematical models according to the grayness of the "black box" problem with which they are faced. Figure 1 shows the spectrum of mathematical modeling problems as they arise in a variety of physical, life, and social science disciplines. Near one end of the spectrum, the "white box" end, we find the mathematical models arising in electric circuit theory. Here one usually knows the structure of the circuit and most, if not all of the element values. Using Kirchhoff's laws and similar network theorems one can construct the mathematical model virtually without recourse to experimental data. Occasionally some parameter values remain to be identified, but this is a relatively simple and straight-forward problem. Proceeding along the spectrum we encounter the problems in mechanics such as aerospace vehicle control. Here most of the model is derived deductively from basic mechanical principles and knowledge of the dimension and characteristics of the system. However, some parameters, for example certain aerodynamic functions, must be identified from actual flight experiments in the presence of noise. Proceeding further away from the white end of the spectrum we encounter the mathematical modeling problems in chemical process control. Here, basic chemical reaction equations and reaction rates are provided. However, a considerable number of variables and parameters are not capable of being directly specified.

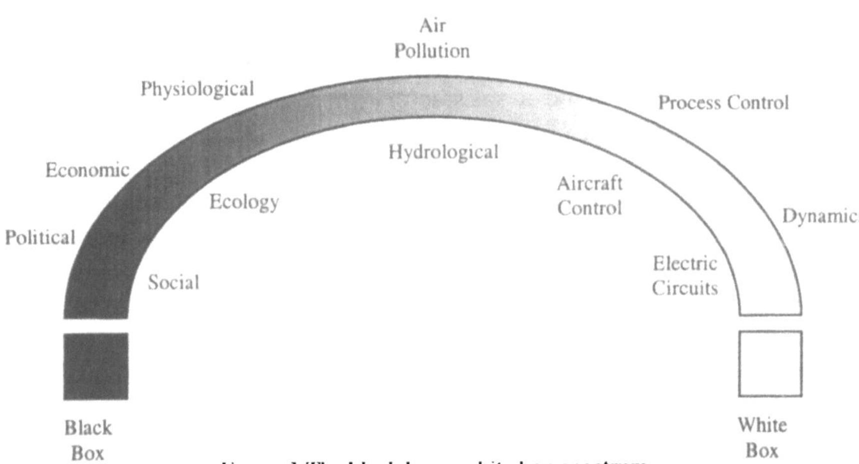

Figure 1 The black box - white box spectrum.

Moving further into the dark area of the spectrum we encounter the models of so called environmental systems. Here there is a general understanding of the physical and chemical processes involved (e.g., movement of water in underground reservoirs, diffusion of air pollutants, etc.). But the field within which these processes occur is not readily accessible to measurements; that is, the phenomena being modeled occur in a medium whose distributed properties are only very imprecisely known. Continuing further into the direction of darkness, a variety of life science models are encountered. Here there is only an approximate understanding of the physical and chemical laws which underlie physiological phenomena, and furthermore the characteristics of the system being modeled are apt to change in time in an unpredictable manner. Economic, social, and political system models fall in the very dark region of the spectrum. Here even the basic laws governing dynamic processes, not to mention the relevant constituents of the system, are open to question and controversy. It is recognized of course that there are many types of mathematical models in use in any specific application area, and that there may well be overlaps in the shades of gray applicable to different fields. The primary purpose of figure 1 is to highlight the existence of a wide range of "gray box" problems, all of which are mathematical modeling problems.

5. Classes of Mathematical Models

All systems exist in a time-space continuum in the sense that inputs and outputs can generally be measured at an infinite number of points in space and at an infinitude of instants of time. Mathematically this means that time and the three space variables can theoretically be considered to constitute continuous independent variables in all system studies. Within the closed region defined by the system boundary, all dependent variables could therefore be expressed as functions of time and three space variables. Since this usually leads to unnecessarily detailed models and to unmanageably complicated equations, the time-space continuum is most often represented by an array of discretely-spaced points in one or more of the four principal coordinates. Attention is then focused upon the magnitudes of the dependent variables at those points rather than at all intermediate points. For example, if the time variable is discretized with a discretization interval of one hour, the system variables are only measured or computed at hourly intervals. Solution values may for example be computed only for noon, 1 PM, 2 PM, etc., and special interpolation techniques must be employed if it is desired to predict the magnitudes of a problem variable at other instants of time. Discretization does not of course imply that the interval between adjacent points in time or space be uniform.

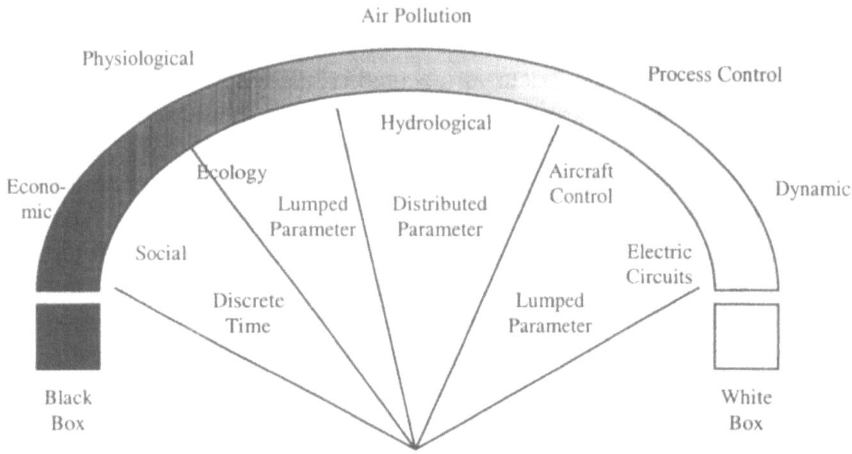

Figure 2. Classes of discretization

In order to classify mathematical models, it is expedient to recognize three broad classes of discretization:

1. Distributed parameter models: All relevant in dependent variables are maintained in continuous form.
2. Lumped parameter models: All space variables are discretized, but the time variable appears in continuous form.
3. Discrete time models: All space and time variables are discretized.

It should be recognized that the above classification refers to the approximations made in deriving the sets of equations characterizing the system. If a digital computer is to be employed to simulate the system or solve the equations comprising the mathematical model, additional discretizations may be required, since digital computers are unable to perform other than arithmetic operations. The three classes described above result in mathematical models comprised respectively of partial differential equations, ordinary differential equations, and algebraic equations. Each of the three classes of mathematical models is employed in distinct regions of the spectrum of mathematical models, as shown in Figure 2. Discretization of space variables involves essentially the representation of the system as an interconnection of two-terminal elements or subsystems. Although each element actually occupies a substantial amount of space, all activities within the element are ignored, and attention is focused only at the two terminals of the element. Such a system is usually represented as a circuit — a collection of elements interconnected in a specified manner. The lines or channels constituting these interconnections are assumed to be "ideal", having themselves no effect upon the system variables other than to act as channels between elements. In order to construct a mathematical model of a system assumed to be comprised entirely of lumped elements, a conservation principle (such as Newton's or Kirchhoff's laws) is employed to provide a separate equation at each of the system nodes — the junction points of two or more elements. For the types of elements occurring in nature, these equations are always either algebraic equations or first or second order differential equations with time as the independent variable. The mathematical model is then a system of equations, one equation per network node. By straightforward techniques, this system of equations can then be converted into a larger system of the first order differential equations, and the dependent variables of these equations are designated as the state variables. The mathematical model of the system is then expressed conveniently in vector form as

$$\dot{y} = f(y, a, u, t), \qquad y(0) = y_0 \qquad \ldots(3)$$

where $\dot{y}$ is the first derivative of the state variable y, a are the parameters and u the inputs. Where the system contains a large number of elements, the state variable vector may become very large, so that the simultaneous solution of all of the state equations becomes excessively time-consuming even where large computers are available. In principle, however, a state variable representation constitutes a convenient and powerful mathematical model. Near the very light end of the spectrum of mathematical models, lumped parameter approximations are used almost exclusively. This is because the systems are actually composed of interconnected components, each capable of being accurately represented as a two terminal element. In making a lumped-element approximation insight into the internal behavior of each element is sacrificed, but in any event this internal behavior is not usually of interest to the system analyst [5-7].

Table 1

	Distributed parameter	Lumped parameter	Discrete time
Type of equation	Partial diff. eq.	Ordinary diff. eq.	Blocks/queues
System constituents	Continuous medium	Circuit elements	Entities
Dependent variables	Potential/flux	Across variables/ through variables	Activities
System parameters	Local field characteristics	Element values	Attributes
Simulation languages	PDEL LEANS	CSMP CSSL Hybrid	GPSS SIMSCRIPT

As one proceeds to the less light regions of the spectrum, into the environmental systems area, it no longer becomes feasible to regard the system as an interconnection of lumped elements. For example, in modeling the atmosphere in an air pollution problem or an underground water reservoir in an aquifer problem the system truly occupies every point in the space continuum, and relevant dynamic processes occur at every point in space and at every instant in time. The space variables must therefore be retained explicitly in formulating the mathematical model. Just as in the case of lumped parameter systems, conservation principles are invoked to permit the derivation of the governing equations. Because of the multiplicity of independent variables, these equations are partial differential equations of the general form

$$\frac{\partial}{\partial t}\left(a\frac{\partial\phi}{\partial t}\right)+b\frac{\partial\phi}{\partial t}=\sum_{k=1}^{m}\frac{\partial}{\partial x_k}\left(\sum_{i=1}^{m}c_{ik}\frac{\partial\phi}{\partial x_i}+d_k\phi\right)+\sum_{k=1}^{m}e_k\frac{\partial\phi}{\partial x_k}+f\phi \quad \ldots (4)$$

Where ϕ is a vector of dependent variables and where the system parameters a, b, c, d, e, f may be functions of the space and time variables as well as of ϕ. It is the uncertainty as to the magnitudes of these parameters throughout the time-space continuum that accounts for the relatively darker shade of gray of the models falling in this region of the spectrum. Often a considerable number of the parameters of eq. (4) may be taken to be zero, so that simplified forms of this equation result. The solution of eq. (4) and most of the simplified forms occurring in the analysis of systems existing in nature, is never easy. Further simplifying assumptions are always required to permit computer implementation, and even then major obstacles arise in the solution of the simplified model [8-10]. As we proceed to the darker regions of the spectrum of Figure 2 into the area of the modeling of biological systems, we find that lumped parameter models again make an appearance. In modeling various phenomena within living organisms as well as in studying the interaction of various organisms in systems ecology, ordinary differential equations are frequently employed. The rationale for employing this class of models is different from that used near the lighter end of the spectrum. In the case of electrical and mechanical circuits, lumped parameter models are indicated because the system itself is in lumped element form, and this type of representation guarantees a highly valid characterization of the system. In the case of biological systems, on the other hand, the system itself is not capable of being subdivided or compartmentalized in such a convenient matter. At the same time there is a lack of basic principles such as Maxwell's equations or the Navier-Stokes equations, which permit the reliable derivation of the basic equations underlying distributed parameter systems. In fact the basic laws are only very incompletely known or understood. Under these conditions the derivation of an equation which would permit the prediction of the system variables at all points in space would be unjustified. In other words, our incomplete understanding of the natural phenomena under study makes it inappropriate to attempt to predict the magnitude of the system variables at all but at a limited number of discretely spaced points within the system. The move from distributed parameter models for environmental systems to lumped parameter models for biological systems is therefore a step in the direction of weaker and less reliable mathematical models [11-13]. As we move into the social science disciplines near the dark end of the spectrum, fundamental laws are almost completely absent; and the elements or subunits of the system under consideration are poorly defined and difficult to describe in terms of input/output behavior. Furthermore, the boundaries between the system and the external world are imprecise or blurred. Under these conditions, deduction from the general to the specific becomes virtually impossible, and major emphasis must be placed upon system observations. Frequently, the causal relationships between variables are poorly understood so that mappings between unrelated variables may be attempted. A great deal of emphasis is thus necessarily placed upon the intuition or insight of the modeler. Sometimes such insight leads to the formulation of lumped models characterized by ordinary differential equations. These models are however recognized as being highly approximate in nature and not to be used for detailed prediction [14]. It therefore becomes expedient in most modeling situations involving such systems to replace the continuous time variable by discretized time. In such a model no attempt is made to describe in detail the dynamic processors occurring between the discretely-spaced points in time. Instead, matter, energy, or information is permitted to accumulate until a designated "event" occurs, at which time this accumulation is reduced. The mathematical models formulated on a discrete time basis are therefore comprised of systems of simultaneous algebraic equations and require the application of queuing theory. Moreover, since the arrival and departure of the matter, energy of information flowing between sub-units is not assumed to occur continuously in time, it is expedient to employ probabilistic measures of arrival and departure rates [15-17]. The three major classes of mathematical models are contrasted in Table 1. Included in this table are the commonly used terms for the description of the variables and parameters of the models. Also included in Table 1 are some of the better known digital simulation languages which have been developed 1 for each of the three classes although hybrid computers have been used from time to time for each of these l classes, their main application has been and continues to be in the lumped parameter area.

6. Validity and Credibility

Validity and credibility are important measures of the quality of a model and of the success of a simulation. Clearly these measures must relate to the objectives of the modeling effort. The motivations for constructing a model and the ultimate use of the model differ markedly for different shades of gray. As one proceeds from the light end of the spectrum to the dark end, there is a gradual but steady shift from the quantitative to the qualitative, as shown in Figure 3. Near the "white box" end of the spectrum, models are an important tool for design. For example in electrical circuit design, models permit experimentation with various combinations of circuit elements to obtain an optimum filter characteristic. Here the validity of the model is such that the errors inherent in modeling can be made small compared to the component tolerances normally associated with electrical circuit elements. Similarly in the area of dynamics, models can be employed to predict to virtually any desired degree of accuracy the response of the system to various excitations. Such quantitatively-oriented models can be used with great assurance for the prediction of system behavior. Closer to the "black box" side of the spectrum, models play an entirely different role. Frequently they are used to provide a general insight into system behavior — behavior which is often "counter-intuitive". Thus, systems containing many complex feedback loops may actually respond to an excitation or control signal in a manner that is diametrically opposite that which was expected. Forrester has pointed out this use of models in connection with simulations of urban systems. Occasionally, the primary objective of the model is to arouse public opinion and promote political action by suggesting that current trends lead to disaster in the not too distant future. Ranged between these two extremes in the motivation for mathematical modeling lies a plethora of part-qualitative and part-quantitative positions. It is very important to recognize, in evaluating and in using mathematical models, that each shade of gray in the spectrum carries with it a built-in "validity factor". The ultimate use of the model must be in conformance with the expected validity of the model. Likewise, the analytical tools used in modeling and in simulation should be of sufficient elegance to do justice to the validity of the model; excess elegance will usually lead to excessively expensive and meaningless computations.

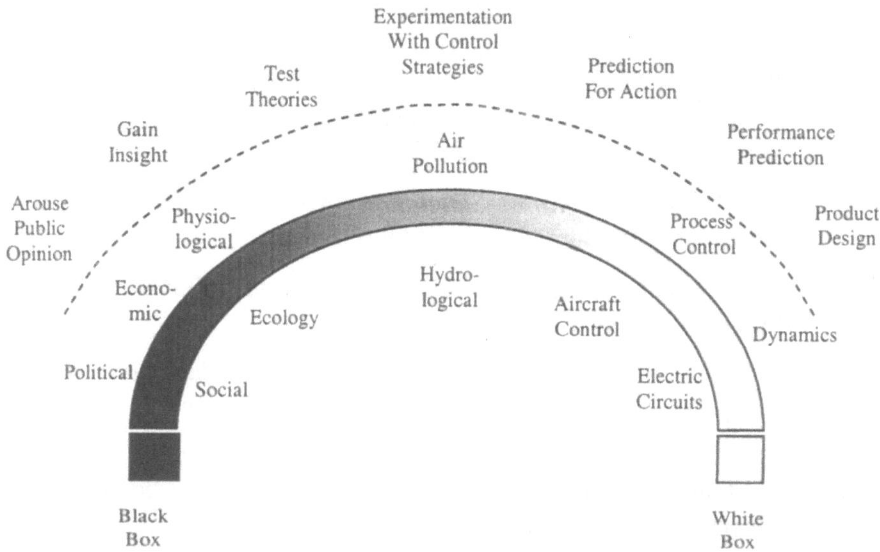

Figure 3. Motivations for modeling.

Serious misunderstandings and disappointments have been caused by a lack of awareness on the part of all concerned with the limitations of mathematical models and simulations. It is of paramount importance that the designers of models and the specialists in the art of simulation communicate to the ultimate users of the models the general validity of the model predictions. It is important moreover that this communication be couched in terms readily understandable by the ultimate user and that they be repeated sufficiently frequently so that no improper inferences be drawn from computer printouts or other computational displays.

References

[1] T.S. Kuhn. *The Structure of Scientific Revolutions.* University of Chicago Press, Chicago, 1962.

[2] D. Bohm. *Causality and Chance in Modern Physics.* Routledge and Kegan Paul Ltd., London, 1957.

[3] M. Born. *Natural Philosophy of Cause and Chance.* Clarendon Press, London, 1949.

[4] P. Frank. *Philosophy of Science.* Prentice-Hall, Engle wood Cliffs, New Jersey, 1957.

[5] Y. Chu. *Digital Simulation of Continuous Systems.* McGraw-Hill Inc., New York, 1969.

[6] *Proc., IBM Scientific Computing Symposium.* "Digital Simulation of Continuous Systems". International Business Machines Corporation, White Plains, New York, 1967.

[7] W. Jentsch. *Digitale Simulation Kontinuierlicher System.* R. Oldenbourg Verlag, Munich, 1969 (German).

[8] W. Karplus. *Analog Simulation: Solution of Field Problems.* McGraw-Hill Inc., New York, 1958.

[9] G. Vansteeenkiste (Editor). *Proc., IFIP Working Conference on Computer Simulation of Water Resources Systems.* North-Holland Publishing, Amsterdam, 1974.

[10] G. Flemming. *Computer Simulation Techniques in Hydrology.* Elsevier, New York, 1975.

[11] D.D. Sworder. "Systems and Simulation in the Service of Society". *Proc. of Simulation Councils*, 1 (1971).

[12] G. Vansteenkiste (Editor). *Proc., IFIP Working Conference on Biosystems Simulation in Water Resources Systems.* North-Holland Publishing, Amsterdam, 1975.

[13] G.S. Innis (Editor). "Simulation Application in System Ecology". *Society for Computer Simulation Proc.*, 5, 1975.

[14] J.W. Forrester. *Urban Dynamics.* MIT Press, Cambridge, Mass, 1969.

[15] G. Gordon. *System Simulation.* Prentice Hall, Engle wood Cliffs, New Jersey, 1969.

[16] H. Maisel and G. Gnugoli. *Simulation of Discrete Stochastic Systems.* Science Research Associates Inc., Chicago, 1972.

[17] P. Rivett. *Principles of Model Building.* John Wiley and Sons, London, 1972

7

Models and Reality: Some Reflections on the Art and Science of Simulation*

George A. Bekey
Department of Computer Science,
University of Southern California
Los Angeles, CA, 90089
bekey@usc.edu

Summary. This paper presents a philosophical overview of simulation, with an emphasis on the nature of the relation between models and the systems they represent. Model-making is shown to be an essential feature of human thinking. The "abstraction ladder," which leads from observations of reality to mathematical models, is reviewed from a semantic viewpoint. The problem of validation of the results of a simulation is then outlined. The paper concludes with same philosophical observations on the relation of model and reality.

1. Modeling as a Fundamental Human Activity.

What is a model anyway? A model has been defined as an idealized representation of the real world; a simplified version of more complex forms, processes, and ideas which may enhance *understanding* and facilitate *prediction*[1]. Note that from this point of view nearly every noun is a model of some event, process or idea. The word *TREE* or the picture in Figure 1 is both models of something in reality. Clearly, neither symbol does a very good job of representing the Canary Island pine in front of my house, but both communicate something to other human beings who associate them with their own experiences.

Tree

(1a)

(1b)

Figure1 – Model of a tree

* Reprinted from Simulation, Vol. 29, pp. 161-164, November 1977 with permission of the publisher

And there lies the rub: semantic symbols (words) or pictorial symbols are only references to experience. Hence the statement *I know what you mean* is probably not verifiable (as discussed by Weaver in an eloquent essay[2]). Verbal descriptions, then, may be models of the relationship between an event and an observer. We shall return to this idea later in a discussion of semantics in Section 3.

But what about modeling and simulation in everyday life? Think about the child who pretends to be an adult by wearing father's or mother's shoes. This is a simulation of adulthood, as are playing house, doll houses, lemonade stands, and countless other childhood activities. Expectant mothers perform simulations in which nursing and diaper changes are practiced on dolls, budding nurses and physicians practice giving injections to oranges, high-school students drive simulators at school and small-scale cars at Disneyland, generals play war games in which pilots fly simulated missions, new paints are tried in simulated environments, and we use group therapy and encounter groups to simulate different human relationships.

If one looks carefully at human activity, one finds that the use of models and the simulation of activity are ever present, from early childhood to old age. It often appears hidden under such phrases as *Let's try the following.* We simulate by actions or by thought experiments. In fact, we simulate so much that the distinction between model and reality may become fuzzy at times.

Ah, and there is the problem! Am I eating imitation meat and drinking imitation orange juice on a table of synthetic wood (patterned like walnut) while my children play on the artificial grass? Our homes are filled with simulated stone and wood, artificial fruit and flowers, simulated fireplaces, simulated brick walls, and simulated masterpieces. And yet we say to one another *Don't take any wooden nickels* – implying that one must carefully distinguish the simulated from the real.

2. Models as Aids to Understanding Reality

An important purpose of models is to help us make sense of complex and baffling events in the world around us. Hence, such models may be highly simplified. A young woman who says *Men are only after my body* is expressing a simple model of masculine behavior. Similar simplistic models of human behavior are embodied in statements like *Homosexuals can't be trusted* or *Russians only understand force* or *All politicians are crooks* or *Unemployed people are just lazy.* Statements like these (and many proverbs in all cultures) represent simplified views of human behavior which enable us to understand, i.e., to make logical, what otherwise would be highly mysterious. Of course, as with other simple models, observation and experience should lead to their modification and improvement.

Models are used in considerably more complex ways to improve understanding. In this class are a Freudian model of human behavior, Toynbee's or Marx's model of history, Darwin's model of development of animal species, the supply-and-demand "law" of economics, or Newton's "laws" in physics. In fact, such model-making is fundamental to the scientific method, since each model is but a hypothesis concerning the behavior of some process. The critical factor arises from the fact that such models have a tendency to become substitutes for reality, i.e., to become reality themselves. The essence of the scientific method lies in the use of models for prediction, and the alteration of the models as our predictions don't quite match observations of the real world.

3. Models and Reality

Models are not the real world. The examples in the preceding sections were intended to illustrate the fact that a model, no matter how complex, is but a representation of reality and should never be confused with it. Golomb[3], in a brilliant article on mathematical models, illustrates this confusion by a man who eats the menu in a restaurant or tries to find oil by drilling through a map of Texas. But how can such confusion arise?

Probably, at least part of the problem is our language, which tends to reduce the ceaseless flux of experience to frozen symbols. While there are "process languages" (Whorf [4]), the majority of western languages are thing-oriented, and thus they concentrate on nouns which freeze experience. The tree outside my window is continually changing, but the word "tree" is a fixed symbol.

Hayakawa[5] illustrates the problem by the ladder of abstraction, which is illustrated in Figure 2. At the bottom of the ladder is a particular brown and white cow which I can see from my window and whom I call Bessie. The next level is represented by the word "cow", which includes Bessie, Elsie and thousands of other unnamed similar creatures. "Mammals" as the next level is even broader and further removed from immediate experience. Finally the word "animal" has very little in common with my experience of Bessie with her big brown eyes.

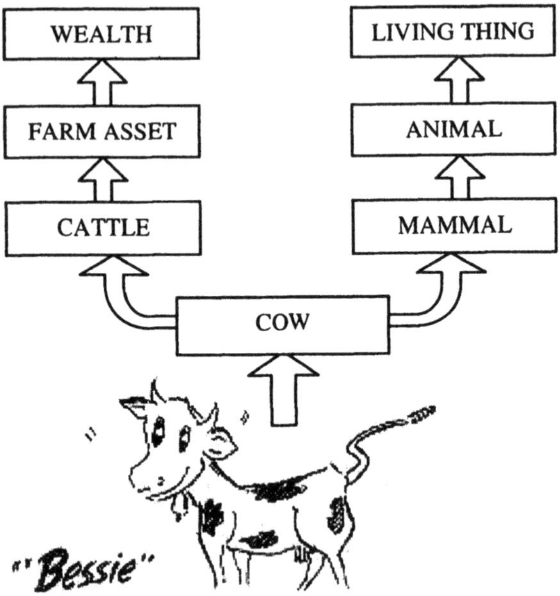

Fig 2: The Ladder of Abstraction

Note that Figure 2 also includes a separate branch of abstractions which leads from Bessie to farm assets. Such abstractions are essential for communication - and many complex models of behavior are based on them. However, they are *not* the real world - and refinement of the models based on experience must be based on observations of Bessie and Elsie and abstracted up the ladder to modify our models. How much easier it is to stay high on the ladder and comment on the behavior of criminals or politicians or cattle as an investment!

4. Models in Science and Engineering

The previous paragraphs have concentrated on models in popular thought and language, to illustrate the fact that the processes of modeling and simulation are ubiquitous. It should be noted that the same problems exist in science and engineering. We tend to accept models derived at some level of the ladder of abstraction and then to "freeze" them into eternal truth. Just think of how hard it is to change scientific ideas like Bohr model of the atom, the Freudian view of the man, or the "big-bang" theory of the origin of the universe. Many scientific issues, such as the fluoridation of the water supply or the use of coal instead of oil, tend to be debated at high levels of the abstraction ladder, in terms of their political, economic, or psychological significance. (See also Siu.[6])

I believe that part of the problem arises from the fact that much of our education is deductively rather than inductively oriented. To teach electromagnetic theory on an inductive basis would require us to move pieces of wire in front of magnets and then try to see whether from these experiments we could somehow synthesize a model. The deductive approach states Maxwell's equations and deduces various consequences from them. In the deductive approach Maxwell's laws tend to assume an almost eternal validity, while in the inductive method they are merely a model to explain some specific and isolated experiments, subject to change as new experiments are performed.

Of course, this is highly oversimplified, but it is based on the observation that many technical and scientific people cling to theories and models as if their lives depended on them. We often use phrases like *Don't knock it; it works!, It would cost too much to change,* or *We always use Method A*. We sometimes distort data to fit straight –line relationships (because it is easier) and throw away points that don't fit the theory. In probabilistic models we look at data through rose-colored glasses so that all distribution look Gaussian. But then, eventually, we develop mathematical models of systems and processes, and solve them on computers. This process we call *computer simulation*, and then we try to make predictions about the actual system behavior from the simulation. And this is easier said then done.

5. The Problem of Model Validation

The validation problem is essentially that of the scientific method:

(1) The model (or hypothesis) is formulated.
(2) Model predictions are tested by comparing them with real world behavior
(3) The model is modified so as to reduce the differences between model and real -world behavior.

Unfortunately, step 2 of this process becomes more and more difficult as the model is further and further removed from the observations (i.e., as it is located further up the abstraction ladder).

Consider, as an example, the computer simulation of helicopter dynamics shown in Figure 3. At the reality level (1) are observations and measurements of airframe and rotor behavior. Collected, these form the data (Stage 2) but they contain the measurement errors due to imperfect instrumentation and the presence of measurement noise, donated by e_M in the diagram. At level 3, a general mathematical model is formulated. Of course, this model will omit those aspects of the helicopter which are not well understood or for which data are difficult or impossible to obtain. These errors in model construction, we term *structural errors* and denote them by e_S. A second phase of structural errors arises from the simplifications introduced into the model of Stage 3, which makes its solution computationally feasible with available equipment. For example, in Stage 3 the poorly understood aerodynamics of rotor downwash may have been neglected, but rotor blade-bending may have been included. At Stage 4 we my neglect all blade-bending and we linearize the equations by restricting certain motions to small perturbations about trim conditions. Now the model of the Stage 4 is programmed, introducing round-off, truncation, and approximation errors, which we collectively call the numerical errors e_N of Stage 5.

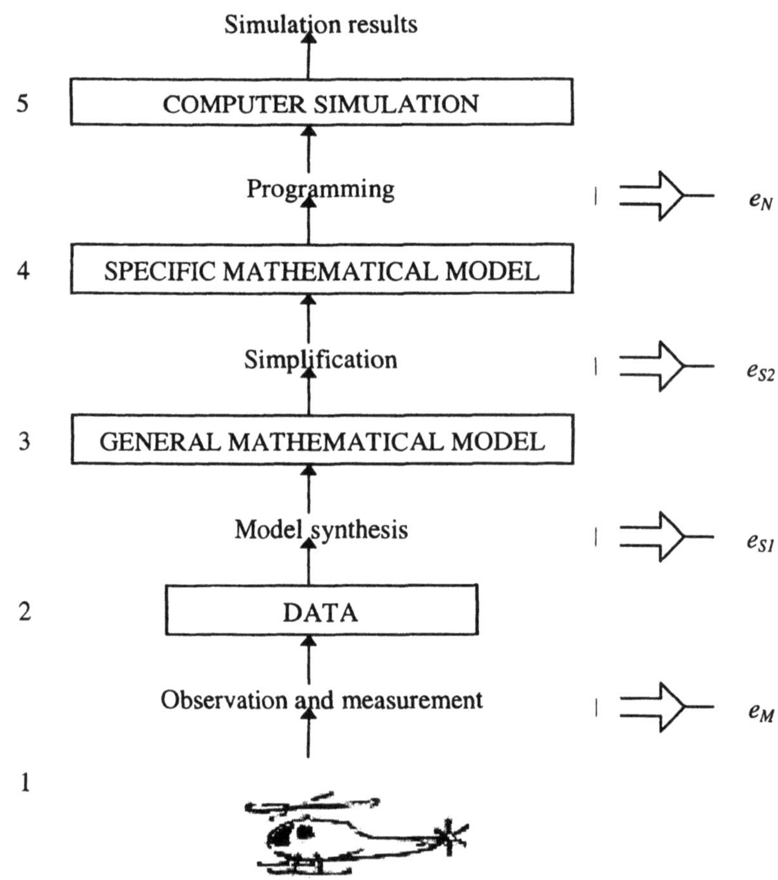

Figure 3- The abstraction ladder to computer simulation

Now it is evident why the problem of model validation is so difficult: we must compare observations at two widely separated levels of the abstraction ladder with one another and, after corrections for measurement and numerical errors, modify the models so that structural errors are reduced. In our experience, this process is really feasible only with relatively simple models. Hence, a common approach is to reduce or simplify the models in Figure 3 until all errors are understood. Then, a difference between model and system behavior can be rationalized and corrected. Finally, the process of synthesis or aggregation of submodels must take place. However, the question of combining experimental errors from a number of nonlinear subsystems is by no means trivial. For a mathematical treatment of this subject, see Reference 9.

Probably it is fair to say that at this time very complex models are simply not validated completely. The process is too difficult. In practice, enough simplified cases are run until either (a) the investigators have developed some of measure of faith in the model (i.e., they are willing to accept it in lieu of reality) or (b) the validation budget is exhausted and results must be obtained.

Finally, it should be noted that there are models which, by their nature, can not be validated exactly. For example, models of the creation of the solar system simply can not be validated exactly, even if model predictions for 1985 match reality to within the bounds of experimental error.

6. Some Further Observations on the Modeling Process.

We have concentrated on the question of relating simulations with the real world because this problem appears to be one of great importance. To reduce the problem, we make the following suggestions:

a) Always remember that simulations are *partial* – they never represent the entire system or process. Hence, keep in mind what has been omitted and what assumptions have been made in model formulation.

b) Models have a specific purpose. To use models for purposes other than those intended and thus to try to validate their results is folly, unless the objective is to modify and improve the models themselves. In addition, we offer the following, paraphrased from Golomb[3]:

c) Do not fall in love with your model.

d) Take all results of your simulations with a grain of salt.

e) Don't distort reality to fit the model.

f) Don't retain a discredited model.

g) Don't extrapolate beyond the region of fit.

In summary: Keep everpresent in your mind the difference between model and reality.

7. A Philosophical Concluding Thought

Models, then, are an essential part of the thought process we call the scientific method, provided that we remember that they are not reality. But this raises a question: *What is reality?* A classical Hindu prayer says *Help us to distinguish the unreal from the real,* which raises the caveat about wooden nickels to the spiritual plane. And interestingly enough, if one looks at the world's religions it is here where the question of models of reality assumes many strange forms. In Hinduism, *Maya* is some times interpreted to mean that the whole world of sensory experience is illusory and that some other experience is real. Probably a better explanation[7] is that excessive identification with the symbols of experience obscures the experience itself-but this only tells us to distinguish the symbol from reality. There is a classical Buddhist painting (one version of which hangs in my office) which shows a monkey reaching for the reflection of the moon in the water. What a clear statement!

In the Judeo-Christian tradition, the making of models of Ultimate Reality is forbidden: *Thou shalt make no graven images . . . .* (Exodus 20:4). But man is a symbol-making animal – and models creep in again and again, sometimes in strange ways. In his recent book, *Simulations of God*, Lilly[8] discusses such models as death, money and humor.

Finally we draw on the Tao-The Ching, where the opening chapter states: *That which can be expressed is not the eternal Tao.* We translate that to mean: That which can be simulated is not the real thing. Keep that in mind, fellow modelers.

Acknowledgement

Preparation of this paper was supported in part by NIH grant # GM 23732 and in part by NSF grant # ENG 74-02297 AOI.

References

[1] M. Rubinstein. *Patterns of Problem Solving.* Prentice Hall, Englewood Cliffs, New Jersey.

[2] C. Shannon and W. Weaver. *The Mathematical Theory of Communications.* University of Illinois Press, 1949

[3] S. Golomb. "Mathematical Models - Uses and Limitations". *Aeronautics and Astronautics*, 57-59, January 1968, Reprinted in Simulation, 197-198, April 1970.

[4] B.L. Whorf. *Language, Thought and Reality.* MIT Press, Cambridge, Massachusetts, 1956, Wiley New York.

[5] S.I. Hayakawa. *Language in Action.* Harcourt, Brace & Company, New York, 1949.

[6] R.G.H. Siu. *The Toa of Science: An Essay on Western Knowledge and eastern Wisdom.* MIT press, Cambridge, Massachusetts, 1957, Wiley New York.

[7] A.K. Coomaraswamy. *Hinduism and Buddhism.* Philosophical Library, New York, 1956.

[8] J.C. Lilly. *Simulations of God.* Simon and Shuster, New York, 1975.

[9] T.M. Grove and G.A. Bekey. "Sensitivity Problems in the Identification of Biological System Parameters". *Proc. Symposium on Applications of Computer Methods in Engineerin,* 625-634, 1977.

8

Structure Characterization of Ill-Defined Systems

G.C. Vansteenkiste
Professor of Engineering
University of Ghent, Ghent, Belgium
ghislain.vansteenkiste@rug.ac.be

J.A. Spriet
University of Leuven, Leuven, Belgium
jan.spriet@agr.kuleuven.ac.be

Summary. Recently, it was found that pattern recognition techniques could be used in the context of model structure identification and hence in the validation stage of the simulation process. The area of life-science systems could benefit from this to create mathematical models showing extended application capabilities.

1. Introduction

This chapter studies the expansion of modeling towards ill-defined systems, more particularly life-science processes. The content was composed from research done in cooperation with Prof. Jan Spriet (University of Leuven, Belgium). Part of the literature review on this topic was composed at U.C.L.A. and many aspects were discussed with Prof. Walter Karplus. His broad scientific view formed an overwhelming source of challenges in the research career of the authors.

2. Mathematical Modeling in Biosystems

In order to make a proper diagnosis concerning the causes of the unsatisfactory achievements of the system engineering approach in life sciences studies, a closer investigation into the relationship between process, scientific knowledge, and modeling methodology is required. Reference is made to "The Spectrum of Mathematical Models" by Prof. Walter Karplus – see *Methodology Section* of this Memorial Volume- and depicted in Figure 1.

One has to admit that engineering activities bear heavily on scientific knowledge. For long times, the scientist has carefully studied natural phenomena. His major concern is to understand unknown processes. The engineer, on the contrary, considers phenomena from another point of view. His objective is to design or modify the processes, so as to make them useful and profitable. To reach this goal, he will often make use of design methodology or system theoretical results for control and prediction. This approach, however, requires a mathematical model. One of the major tasks of the engineer, therefore, consists of constructing a mathematical representation which should be a valid abstraction of the underlying phenomena. As the engineering system is new and original, some form of creativity and skill is required to obtain the model. Nevertheless, the model builder will rely on the associated scientific discipline to minimize his effort. It is rarely taken into account that the knowledge available in the field under study has never a uniform structure: certain statements have more validity and generality than others.

All scientific endeavor starts with fact finding, observations and experimentation, often followed by curve fitting and derivation of simple mathematical relations. Taking into consideration the difficulties and uncertainties of that initial work, there is no doubt that most of the data and facts have low validity and narrow generality. As more evidence is gathered, one is able to make assumptions, to draw conclusions, to posit hypotheses and to generate small-scale models. Some deeper understanding is gained and pieces of information are available that may be considered of medium validity and limited generality. The most difficult but most gratifying step is to assemble all pieces into a general theory. The principles, laws and relations that form the theory have then necessarily a broad generality and high validity. As the theory is able to explain a wide variety of phenomena, it is considered to be the truth at a certain point in time. The model builder who tries to solve his particular problem can draw a priori knowledge from any of these categories along route I, II or III on Figure 1

Mathematical models are generated in part by application of a priori knowledge which should have high validity and therefore should be drawn along route I. As bioprocesses will include physical phenomena, the general theory explaining these phenomena can be used. For the typical biological aspects, however, the general theory is not so well developed, it has less validity, and it is seldom expressed in mathematical terms. Consequently, the engineer has to use routes II and even III to assemble the necessary a priori information, and that information is unfortunately of more dubious validity.

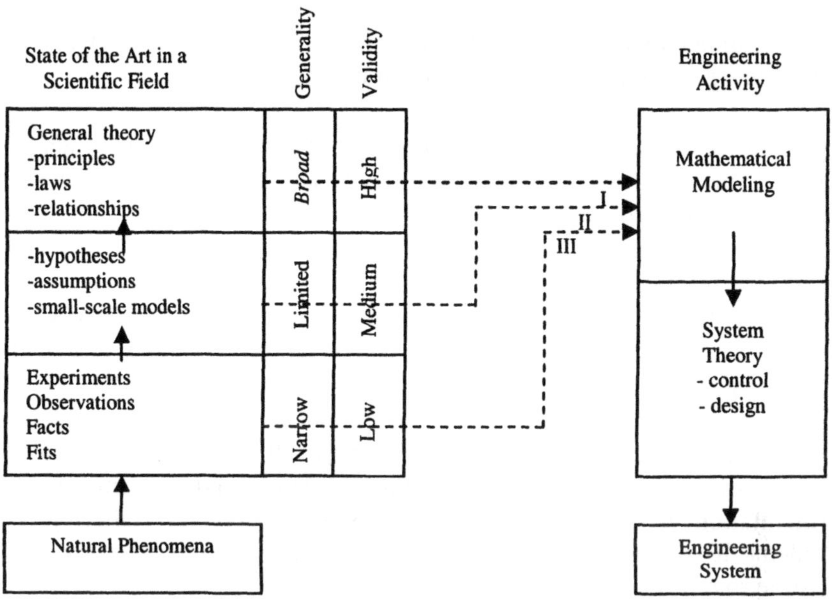

Figure 1. Science-Engineering Interaction

To illustrate the deficiencies in the general theory (the so-called softness or grayness), a few examples are given. Species interaction is of key importance in the study of ecological systems. According to Darwin, interactions are dictated by the law of survival of the fittest while changes in genetic structure occur at random. The law is widely accepted, but is in a verbal form. Furthermore, details are missing, and it is impossible to give the law a precise mathematical representation. At last, although the theory has been considered for a long time as having broad generality and high validity, more and more biologists start to question the usefulness, the relevance, and even the trustworthiness of the theory [1]. Another important principle in biology is the fact that organisms may be divided into classes depending on morphology, growth form, and their mode of reproduction [2]. Here the validity is well established, but again the rule is qualitative and its impact on models is poorly understood. In short, one has to conclude that the theoretical foundations in biology have not yet reached the level of abstraction of certain other fields. The deficiency in scientific background is a first obstacle and consequently, very often the methodology of the systems analyst will produce less valid representations.

Biosystems themselves have number of properties which make a quantitative approach cumbersome and difficult to carry out in a rigorous way. Ill-definition is a generic term for these undesirable properties.

First the processes are often of an intricate nature. On the average, the number of meaningful interconnections will be larger than in physical disciplines. A cell, for instance, has a large number of highly integrated subsystems.

As a second feature, it can be mentioned that biological systems are as difficult to separate in space as in scale, and often in time. It is one of the major principles in biology that the activities of an organism and the rates at which it carries on those activities are dependent not only on the organism itself but also on the organism's environment [2]. It is therefore difficult to define clearly the boundaries of the system, and separation for experimentation is almost unfeasible without substantially altering the phenomena. It is, for instance, well known that in vivo processes are often impossible to reproduce in vitro.

Difficult spatial separability is not the only hindrance. Scale differences are also small, so that it is sometimes not clear which level is appropriate. Again, comparison with physical phenomena is helpful. For electro-magnetism, e.g. a microscopic theory is available that explains the phenomena at the atomic or subatomic level. There exists also a macroscopic theory, represented by the Maxwell equations, which describes larger objects. For the latter theory, microscopic events have been averaged over an appropriate region. Scale differences between microscopic entities : electrons, nuclei, etc. $\left(< 1\text{Å}\right)$ and the macroscopic systems $\left(> 10^8\,\text{Å}\right)$ are impressive and they explain, for a part, the extremely accurate results obtained for such systems. Chemical models are another example. Chemical concentrations assume a large number of molecules within a tiny space and a macroscopic point of view is acceptable.

However, if one considers size differences between an organism $\left(10^8 - 10^9\,\text{Å}\right)$, a cell $\left(10^4 - 10^5\,\text{Å}\right)$, and a macromolecule $\left(10^2 - 10^3\,\text{Å}\right)$, it is seen that the different hierarchical levels are less far apart from each other, and determination of an appropriate level is consequently less evident. For a population of cells, for example, it is difficult to decide whether modeling at the cell level or a macroscopic population-oriented approach is better adequate for a particular purpose. This fundamental question is still largely unsolved. A beautiful illustrative example is the state-of-the-art in modeling biomass for fermentation processes. A classification of mathematical models of microbial populations has been made by Tsuchiya, Fredrickson and Aris [3], see Figure 2. Segregated models consider individual cells, and therefore start from a lower scale level than non-segregated models where the biophase has been lumped through averaging. In both categories, the biophase can be given some structure, but this is a matter of modeling detail and aggregation. The first classification is more fundamental. The major problem remains as to indicate which models are adequate for given engineering problems.

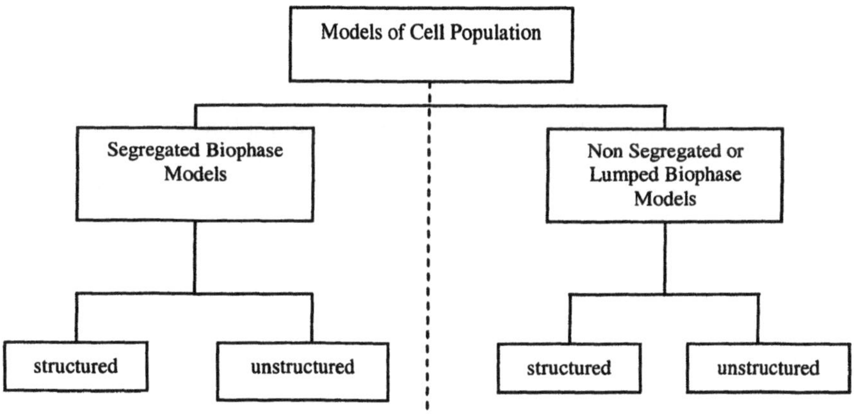

Figure 2. Models of Cell Population (Tsuchiya, Frederickson, Aris)

Beside scale differences, time separation may also be difficult. In electromagnetic theory, a quasi-stationary theory permits easy analysis for relatively "slow" phenomena. Electrical eigen frequencies are much higher than "surrounding" mechanical, chemical or thermal frequencies. In biology, at the contrary, all time constants are usually of the same order of magnitude which makes the separation difficult. Moreover, time constants inherent in biological systems are often comparable to those characterizing the life of human beings which then makes experimentation extremely difficult.

Finally, a low accessibility to the biosystems has to be recognized. For almost all biological processes, experimentation is a nuisance not only because of the processes being difficult to separate, but also because of measurement techniques being inadequate and/or very expensive. Some facets of low accessibility can be summarized as follows:

- Many measurement techniques are based on chemical analysis. Such an approach requires a lot of manipulative action and is prone to a wide variety of biases and errors.
- In the case of animals, ethical considerations may set limitations on experimental set-ups and for human beings those restrictions are even much more severe.
- The inherent variability between different organisms of the same kind limits drastically the advantages of consecutive or parallel experiments.

All these aspects of biosystems which make them ill-defined, form the second important stumbling block for systematic and valid modeling.

The final step in reaching an engineering goal is the construction of a specific mathematical representation and the application of system theoretical results. Historically, system analysis originated from pure mathematics, and the theory has been built up around linear systems because of their analytic tractability. It happens to be that the tools developed were adequate for the study of many physical processes. Biosystems, on the contrary, are highly non-linear and adaptive. That latter feature is even one of the major principles in biology, and it is usually stated as follows: The current phenotype (state or constitution) of an organism depends not only on its genotype but also on the past history of environments seen by that organism. Mathematical formulations with such properties are not always well understood. Another aspect of biological entities is their variability. This can be taken into account by postulating stochastic models, the theoretical foundations of which are (even in the non-linear case) yet better established. Practical application methodology is, however, lacking. The Monte-Carlo approach, a brute force technique, is still the only general method available.

The modeling methodology that has been advocated consists of a largely deductive approach forming an interface between a priori knowledge and the parameter estimation or calibration and validation stages as the two major and most important inductive steps in the overall procedure. The techniques for these latter two steps often assume large data sets. This global approach in fact assumes implicitly that the a priori knowledge is of good quality and validity and therefore, (referring to Figure 1) that the knowledge is drawn along route I from a comprehensive and powerful general theory. Furthermore, the iterative loop requires the possibility of easy experimentation. It is clear that the properties of softness and ill-definition violate the implicit assumptions present in the modeling procedure.

Summarizing, it can be stated that there are three "weak" points in Figure 1: the properties of biosystems, the non-availability of a comprehensive general theory, and the badly-adapted engineering methodology itself. The analysis made can be underpinned with other arguments and facts and, as mentioned above, a certain consensus has grown among modeling specialists about these major causes of the limited success of mathematical models in biosystems.

3. Model Structure Determination

The methodology for structure determination is still in its infancy. The problem itself can be seen as a hierarchical decision tree [4, 5]. In a first stage called model formulation, important factors, input output quantities and state variables have to be selected. In a second stage, the transformations between these variables have to be chosen, except for a number of unknown parameters that remain to be determined afterwards. There exist guidelines, that can help the model builder during that phase of the modeling effort. The structure should be as parsimonious as possible, identifiable and objective oriented; the more physical, the better. Unfortunately these principles are quite general and may be conflicting, so they are of few direct help.

If the system is known to be linear and the important variables have been determined out of the a priori knowledge, the structure characterization boils down to be a problem of determination of the order of the system. Some theories are yet available for this case. Akaike [6] and Rissanen [7] adopt an information-theoretic point of view, while Kashyap [8] stresses a Bayesian approach. For non-linear structures there is a distinct lack of techniques. However a few promising ideas seems to stimulate research.

Each parameter identification yields residuals, which were statistically analyzed. Whenever, these residuals were sufficiently white, the structure was supposed to be correct. If trends could be detected a structure modification was thought necessary. An elegant feature of that approach was that estimation and characterization were performed in one single effort. Drawbacks are evident: bad initial structure will cause long iterations and optimal estimation strategies are likely to be bad structure analyzers. However an important conclusion remains that parameter estimation in a more general sense can be a tool for structure investigations.

A first approach to non-linear structure identification is the use of recursive parameter estimation schemes, Young [9]; the method is a direct extension of the classic approach. One starts with a linear structure and uses the recursive estimations of the parameters as indicators for the structure. Because the recursivity adds an extra-dimension to the estimates, it is believed that additional information is available about structure characteristics. A pleasing feature is the fact that one only introduces non-linearity when it is really necessary, so the principles of parsimony and identifiability are taken into consideration. A doubt, however, remains about the possibility to get from linear configurations to complex non-linear ones which may be more physical in nature.

A second approach relies heavily on probability theory and Bayesian methods and has been introduced by Kashyap [10] A set of candidate structures has to be chosen. Given the data, one looks for the structure class, with the highest a posteriori probability. To some degree, parsimony and identifiability as well as physicality and objectivity can be accounted for, but there is a restriction on the non-linearities. Furthermore, to obtain the a posteriori probabilities, quite a number of assumptions have to be made and the probability densities have to be cooperative. Investigations to test the robustness of the technique when certain assumptions are invalid is certainly necessary.

A third method is known as the 'Group Method of Data Handling' introduced by Ivakhnenko [11]. It is based on the principles of heuristic self-organization and relies on bioengineering concepts. As the brain is a neural network, capable of learning and performing complex control and recognition tasks, it is stipulated that an intricate, non-linear, layered network should be able to simulate a system structure if it is appropriately trained by data. The method seems to negate the principle of parsimony and physicality is found less important than objectivity. It may however be that, in such a network, information is processed in a more optimal way.

Another approach sets the problem in a pattern recognition framework. Structures can be considered as different patterns and characterization may be seen as a pattern recognition task using experience and present information. These main ideas have been set forth by Karplus [12] and Saridis [13]. Simundich [14] and Hofstadter [15] have worked out certain details of the procedure. The present work is an extension of Simundich's results for ordinary differential equations. The framework is quite general. Parametric as well as non-parametric methods can be included and the importance of a learning phase is stressed. The authors believe that after proper formulation, the other approaches, discussed above, may be fitted into the pattern recognition methodology, so that pattern recognition can become the unifying framework for structure characterization.

4. Pattern Recognition for Structure Identification

An ideal situation would be one where the only input to a computer is the data measured from the phenomenon to be modelled and from this the machine would produce a complete model. This is however utopic. A solution has been achieved where one more input is required: an input considered as a library of models. The machine will search among these candidates in order to find a model whose structure is best adapted to the data. The situation that no reasonably good model can be found in the "library" is also possible. Then the operator has to reconsider the problem in order to add new possibilities to the library. The designer has to specify several distinct classes in such a way that later the attribution for the process to a class answers his questions about the characteristics of the model he is looking for.

4.1. Feature Extraction

The most crucial part of the method is one of feature extraction. The choice among the different models will be done on the basis of characteristic expressions. The most absolute freedom exists about how to select these expressions. As always, freedom is a nice thing but very difficult to handle. Some experience is needed to be able to develop good features. They must be rather insensitive to noise and also have good discriminating properties. The number of these features is also free. Very often one feature per proposed model is used, but sometimes a better discrimination may be obtained with more. In other cases not all features are really important. For computational reasons the number of features should not be chosen too high.

In the example that follows, some possibilities will be discussed. It is very hard to give a general rule. Usually the features bear some relationship to the variability of certain parameters in the model, but that is definitely not a necessary relationship.

The different features form a so-called feature space. It is very important that one has a good feeling of what the feature space stands for, since it is in this space that most of the characterization operations are performed.

Computation of the features can be considered as an operation on the set of data coming from a candidate model during the initial phase and from the experiment later on. It should be noted that in this stage, during the initial phase, one learns the operations to perform later on the data, in order to answer specific questions about the character of the tentative model. We assume that the data are available in discrete form; we thus have a set of couples (t_i, x_i) where t_i stands for the independent variable (time instant on which the measurement was taken), and x_i (which may be vector-valued) are the data obtained from the process under study. The experiment is equivalent to a set of such couples

$$D = \{(t_i, x_i) \mid i = 1, 2, 3, \ldots n\}$$

Called the data set. Obtaining the features can be considered as a mapping from the set D of all possible data sets to the feature space

$$\Theta : D \rightarrow R^k : D \rightarrow f$$

Where k is the number of features and f is a point in the feature space. Often, the mapping Θ is multiple valued, that is, more than one point f may be the image of a set D.

This mapping, that is, the procedure for computing points in the feature space from data, is used as a building block for the structure identification algorithm.

From a pragmatic point of view, a given structure results in a set of properties shared by the data generated by that structure. Certain properties are parameter invariant, others are not. Most properties are common to many structures. For a discriminating study, however, it is sufficient to choose for each structure a property, which is sufficiently parameter invariant, and which is not shared, to the same extent, with the other candidate structures. The feature extractor is a set of algorithms that test for each property so that a proper choice be made, afterwards. There are quite a number of ways to construct feature extractors, the procedure followed in [16] will be explained.

It is assumed that a priori knowledge and prior modeling effort have yielded two sets:

- a set of structures V_c, that are to be tested, with their parametric representations $\underline{F}^i$

$$V_c \equiv \{V_i\} \quad i \in I_c \quad I_c \equiv \{1, 2, \ldots, c\}$$

$$\underline{F}^i\left(\underline{\dot{y}}^i, \underline{y}^i, \underline{p}^i, t\right) = 0 \qquad \ldots(1)$$

$\underline{y}^i\,(nx1)$ observation vector

$\underline{u}^i\,(mx1)$ input vector

$\underline{p}^i\,(rx1)$ parameter vector

$\underline{\dot{y}}^i\,(nx1)$ derivatives of $\underline{y}^i$

- a set of data
  $$V_D \equiv \{(\underline{y}_k, \underline{\dot{y}}_k, \underline{u}_k, t_k)\} \quad k \in I_D \quad I_D \equiv \{1,2,\ldots,d\}$$

For noiseless data, $\underline{p}^i$ may be computed out of equations (1) using a few data points. Under the hypothesis of data coming from that structure, the scatter of many parameter evaluations should be zero and different from zero for data coming from other structures. Direct feature extraction uses that property. So it is easily understood that an implementation can have the block diagram shown on Figure 3. The different parts are :

- a data selector : the algorithm generates a set W_D of subsets from V_D, each subset capable of parameter extraction
  $$W_D = \{W_1\} \quad W_1 = \{(\underline{y}_j, \underline{\dot{y}}_j, \underline{u}_j, t_j)\}$$
  $$1 \in I_W \quad j \in I_L \quad I_L \subset I_D$$
- a parameter extractor : these algorithms solve the non-linear algebraic equations (1) for each class separately. So after extraction c sets of parameter evaluations are obtained :
  $$P^i = \{p_1^i\} \quad 1 \in I_W$$
  In fact, the extractor may be, for convenience slightly modified : each linear or non linear operation $\underline{q} = \underline{h}(\underline{p})$, yields quantities $\underline{q}$ that are also suited for the present purpose. A good choice may facilitate parameter extraction or give more noise-insensitive invariants.
- a data concentrator : this unit performs an operation on P^i and should measure the variability within each set. The result is a pattern vector $\underline{\alpha}$ of dimension c.

One realizes that for noiseless data, coming from the ith structure, the ith entry in the $\underline{\alpha}$ vector will be zero, while positive and hopefully large for the other entries. At that point, choosing the structure boils down to finding the smallest entry in the $\underline{\alpha}$ vector. There are however a number of factors that will obliterate the performance of the scheme:

- computational errors will be present
- derivatives are usually not measured and have to be estimated from the data
- noise, that always corrupts the measurements, will influence the scheme.

As long as the pattern recognizer is able to discriminate variation due to those causes, from these resulting from structure differences, the scheme will be of valuable help. Reference [16] elaborates on different choices of operations for the three blocks in Figure 3.

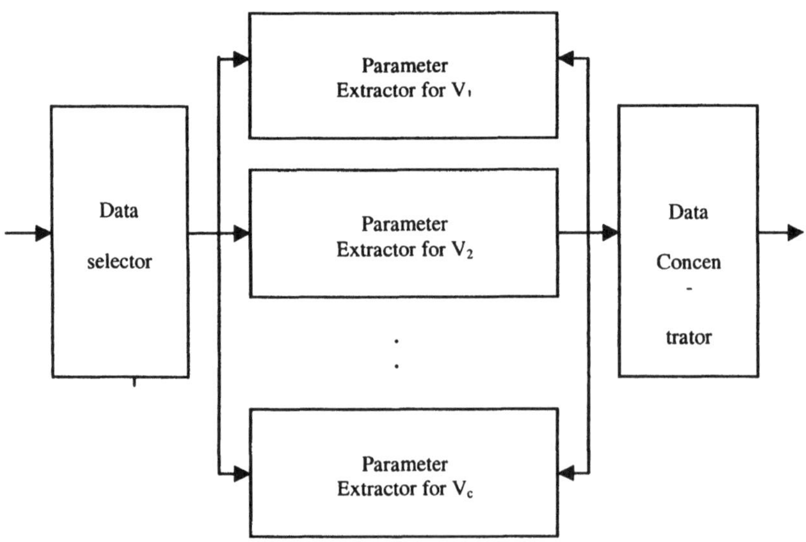

Figure 3. Feature extractor for structure characterization

4.2 Structure Discrimination

Two stages of operation must be distinguished: training of the classification algorithm using the candidate models and using the classifier on the experimental data.

The classifier splits up the feature space in a partition; each subset corresponds to a cluster of points and hence to a model from the library. Training consists of choosing the boundaries in such a way that the partition corresponds to the different models in an optimal way. It is thus necessary to make some simulation runs with each of the proposed models. The data sets from these runs are processed by the feature extractor and the classifier output is compared to the model from which the data were generated. The parameters of the classifier are adjusted in such a way that maximal correspondence between input and output is obtained. The more runs that are used to train the classifier, the better. It is also important to make the different runs of one model as diverse as possible: that is, combinations of parameter values should be used that cover a wide region, so that the set of values that may be expected from real life data, falls inside this region.

4.3. Some Additional Remarks

The user has to keep a close eye on what is happening in the feature space during training. Here, his choice of the computation rules for the features is subject to a very severe test. It may appear that some features are almost useless, or that some kind of modification would entail a considerable improvement in the classification.

It is also during the training phase that the influence of noise on the data must be studied. Often a feature has a very specific behavior for data sets generated with one particular model (e.g. is a constant). When noise is present, deviations will occur, and if the feature computation algorithm is ill-conditioned, these deviations will become so important as to make that feature useless.

Noise can originate from different sources. Not only measurement noise can be present, but noise-like errors may occur in the computation itself. Indeed, often the values of the derivative of a system variable are required. If a derivative is to be computed from sampled data, the exact value will not be found. The situation gets much worse for higher order derivatives; use of these is not recommended.

Training the system requires quite a lot of computation time. This must however be done only once, before the actual experiment start. Batch processing is possible. The computations during the experiment itself are minimal, so on-line computations can easily be performed. An added advantage is that one has permanent control over the model structure. No time will be lost in identifying the parameters of a model whose structure does not correspond to reality.

Even when the model is in use, it is always possible to keep checking that the model structure is still adequate for the incoming data. When deviations occur, it is easy to switch to another model in the library. In this way one will always use that model that is best adapted to the data.

5. Implementation of the Technique to Growth-Model Structures

To test the developed methodology, a process was chosen where intensive modeling efforts provides different valid mathematical structures which are all quite similar. The growth of chemo-organotrophs under a limiting substrate was chosen, displaying a wide variety of organisms and consequently of growth patterns.

From a priori information, it is known that for a simple macroscopic model, biomass x, and limiting substrate S are the important quantities. The mass balance for substrate exchange between micro-organism and environment is given by equation (2)

$$\frac{dS}{dt} = -\frac{1}{y}\frac{dx}{dt} \qquad S(t_0) = S_0 \qquad \ldots(2)$$

The biomass itself changes according to equation (3)

$$\frac{dx}{dt} = f(x,S) = \mu(x,S).x \qquad x(t_0) = x_0 \qquad \ldots(3)$$

The candidate models differ in the form of the growth rate $\mu(x,S)$. Five forms are distinguished, and listed in Table I. The name of the scientist, who first proposed the model, is mentioned. The proposed models entail different physical assumptions.

In contrast with this process, in practice the structure can often not be analytically analyzed, in which case the pattern recognition technique permits a simulation approach to structure analysis. Hence the process allowed the simulation results to be verified.

In reference [16] alternatives are investigated for the processing stages in the different building blocks of a pattern recognizer.

Monod	$\mu(x,S)=\mu_M.S/(k_M+S)$
Contois	$\mu(x,S)=\mu_C.S/(k_C.x+S)$
Verhulst	$\mu(x,S)=\mu_V.(1-x/x_M),x_M=x_0+y_VS_0$
Piece-wise linear	$\mu(x,S)=\mu_P \quad S>K_P$ $\mu_P.S/K_P \quad S\leq K_P$
Tessier	$\mu(x,S)=\mu_T.(1-e^{-S/k_T})$

Table I: Rate equations

A new and very promising method for structure characterization has been explained. Although very heavy computations are required before one can start with actual experiments, the computation load per data-set to be processed is very low. The initial computations can be performed off-line, and hence are less costly than those for the usual real time parameter identification methods. It is our belief that the economic point E (Figure 4) is situated quite far to the left, so that as soon as a few runs have to be identified, this method will be advantageous.

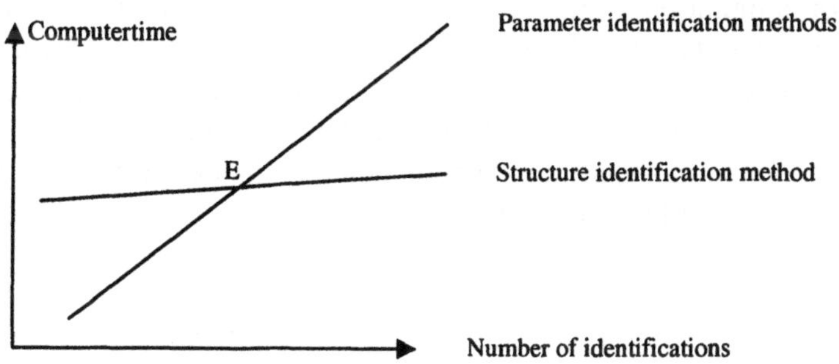

Figure 4. Computation time of the proposed identification method

6. Advances on Important Stages of the Modeling Process

The amount of study, analysis and investigation of techniques for the modeling process has been very unequally spread over the different stages.

An overview of the efforts done is summarized in Figure 5. The methodology of deductive reasoning is well known and established. For the validation step, standard statistical results are available; sometimes the stage may be bypassed and the model is then tested immediately against its requirements according to an evident belief that "the proof of the pudding is in the eating". Further study of the validation techniques in relationship to other modeling stages is still useful. Clearly, most attention has been paid to the parameter estimation problems while other (for biosystems important) issues have been neglected.

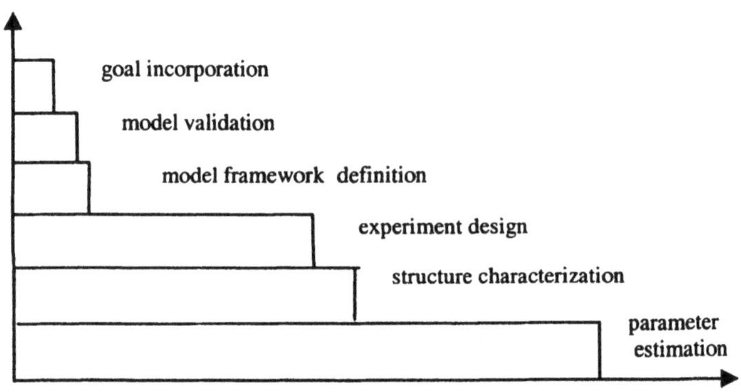

Figure 5. Research efforts in techniques for the different modeling stages

7. Impact on Developments in Simulator Architecture

Ill-defined system modeling, is influencing more and more the architectural design of simulation tools. The current available simulators provide not enough transparency for the user who wants to model ill-defined problems. There is, in other words, too much concentration at the moment on the mechanics of building a sophisticated computer simulation model, and not enough on ensuring that the "sophisticated" model has been adequately identified, estimated and validated.

It has become clear in the previous section that in an experimental study by simulation, the creation and testing of an extensive variety of candidate model structures is of key importance. As a consequence, a simulator for ill-defined systems must be able not only to implement and update data bases, but also to utilize the data base for modeling in a methodology supported way [17]. Simulation methodology should provide algorithms among others for:

- Selecting from the class of available models some candidates having features relevant to a certain experiment; the model candidates may originate from a priori studies or they may come from automated model aggregation in relation to the features present in the available data as e.g. generation of linearized models, sensitivity models [18], meta-models [19], etc.

- Validating a candidate model according to the particular experiment (or vice versa) when the candidate model did not belong to the considered class; some aspects of validation are found in references [18, 20, 21].

Modeling and validation take place on experimental data and general knowledge from the actual system collected and stored in the data base. After the validation process, the experimenter may decide to continue the simulation study by further execution of simulation experiments eventually followed by another modeling phase. As a consequence, a simulation study may lead to the necessity of performing modified experiments on the real process in order to get better or more relevant experimental data. In the future simulator's architecture, data bases will have a key position. Simulators for ill-defined systems should have many features in common with data base machines [22].

Present data base management systems however are oriented towards use by professional programmers, and toward relatively short applications programs which access few records and are run repetitively.

In order to support such areas as the modeling of ill-defined systems, a broader orientation is needed where new classes of users can directly access their data and large collections of data can be manipulated efficiently. Current research activities in this domain are again of the utmost importance for ill-defined modeling support [23] Mathematical modeling becomes really a task which does not start from scratch every time the modeling activity is re-initiated, but is continued from the stored information obtained during previous runs.

8. Conclusion

This chapter has shown that the special characteristics of biosystems have given rise to a more detailed study of the important issues in modeling methodology and tools. As a consequence, mathematical modeling is evolving from an art to a science in the hope that a more systematic approach will produce better performance of the mathematical descriptions and/or will unveil their basic limitations.

References

[1] C.H. Waddington (ed.). *Towards a Theoretical Biology.* Edinburgh 1970.

[2] A.G. Fredickson, R.D. Megee III and H.M. Tsuchiya. "Mathematical Models for Fermentation Processes". *Advances in Applied Microbiology* (Perlman, ed.), Academic Press, 13, 1970.

[3] H.M. Tsuchiya, A.G. Fredrickson and R. Aris. *Advances in Chemical Engineering.* 6:125, 1966.

[4] R. Mehra. "A Summary of Time Series Modeling and Forecasting Methodology". *Proc. of the IFIP Working Conference on Modeling, Identification and Control in Environmental Systems.* Ghent 1977, North Holland Publ. Co, 1978.

[5] H. Akaike. "Stochastic Theory of Minimal Realization". *IEEE T-AC, AC-19*, 6:667-674, 1974.

[6] H. Akaike. "Canonical Correlation Analysis of Time Series and the use of an Information Criterion". *System Identification, Advances and Case Studies,* Academic Press, New York, 1976.

[7] J. Rissanen. "Minimax Entropy Estimation of Models for Vector Processes". *System Identification, Advances and Case Studies,* Academic Press, New York, 1976.

[8] R.L. Kashyap and A.R. Rao. "Dynamic Stochastic Models from Empirical Data". *Math. in Sci. and Eng.*, Academic Press, New York and London, 122: 182-183, 1976.

[9] P. Young and J.M. Kaldor. "Recursive Estimation: A Methodological Tool for Investigating Climatic Change". *CRES Report,* AS/R14, 1977.

[10] R.L. Kashyap. "A Bayesian Comparison of Different Classes of Dynamic Models using Empirical Data". *IEEE T-AC,* AC-22, 5:715-727, 1977.

[11] A.G. Ivakhenko. "The Group Method of Data Handling". *Soviet Automatic Control,* B-3, 43-55, 1966.

[12] W.J. Karplus. "System Identification and Simulation: A Pattern Recognition Approach". *Proc. Fall Joint Comp. Conf.*, 385-392, 1972.

[13] G.N. Saridis and R.F. Hofstadter. "A Pattern Recognition Approach to the Classification of Non-linear Systems". *IEEE Trans. On Systems, Man and Cybernetics,* SMC-4:362-370, 1974.

[14] T.M. Simundich. "System Characterization : a Pattern Recognition Approach". PhD Thesis, UCLA, School of Engineering and Applied Sciences, Los Angeles, 1975.

[15] R.F. Hofstadter. "A Pattern Recognition Approach to the Classification of Stochastic Non-linear Systems". PhD Dissertation, Purdue, Lafayette, Ind., 1974.

[16] J. Spriet and G.C. Vansteenkiste. "Design of a Preprocessor for a Pattern Recognizer Used for the Identification of Soft Systems". *Proc. of the Summer Computer Simulation Conference*, Newport Beach, 1978.

[17] C.R. Standridge and A.A.B. Pritsker. "Using Database Capabilities in Simulation". *Progress in Modeling and Simulation* (F.E. Cellier, ed.), Academic Press, London, 1982.

[18] F.E. Cellier and A. Fischlin. "Computer-Assisted Modeling of Ill-Defined Systems". *Proc. of the 5th European Meeting on Cybernetics and Systems Research*, (R. Trappl, ed.), Vienna, Austria, 1980.

[19] J.P.C. Kleijnen. "Experimentation with Models: Statistical Design and Analysis Techniques". *Progress in Modeling and Simulation* (F.E. Cellier, ed.), Academic Press, London, 1982.

[20] T.I. Oren. "Computer-Aided Modeling Systems". *Progress in Modeling and Simulation* (F.E. Cellier, ed.), Academic Press, London, 1982.

[21] R.G. Sargent. "Verification and Validation of Simulation Models". *Progress in Modeling and Simulation* (F.E. Cellier, ed.), Academic Press, London, 1982.

[22] L. Kunii, Tosiyasu and S. Kunii, Hideko. "A Virtual Information System for Environmental Applications". *Proc. of the IFIP Working Conference on Modeling and Environmental Systems*, Tokyo, Japan, 1976.

[23] *IEEE Transaction on Computer,* Special issue on Data Base Machines, C12, 3, 1979.

9

Inverse Problems

V. Rao Vemuri
Professor of Applied Science
University of California, Davis, CA 94550
rvemuri@ucdavis.edu

Summary. Three inverse problems arising in practical applications are presented. The first problem pertains to the modeling of San Fernando Valley groundwater basin and the primary issue is parameter identification in a quasi-linear parabolic partial differential equation. The second problem addresses the question of determining the optimal number and location of wells to pump, clean up and recharge a contaminated aquifer near the city of Livermore, CA. Results pertaining to these two problems have already been published. In contrast, the third problem, still in a formulation stage, addresses the question of chemical process identification; that is, the task of determining what a chemical factory is producing using incomplete and uncertain information gleaned from photographs, type and quantity of raw materials purchased, effluents produced, the basic laws of chemistry, materials science and common engineering practices.

1. Introduction

Inverse problems frequently arise in experimental situations when one is interested in a description of the internal structure of a system given indirect noisy data. Estimating the response of a system given a complete specification of the internal structure, on the other hand, is the *direct* or *forward* problem.

There are several flavors to an inverse problem. Perhaps the simplest among them arises when one has a mathematical description of the internal structure (typically in the form of an equation along with any auxiliary conditions and constraints) and the task is only to estimate the values of the unknown parameters. This is the *parameter estimation* problem. A somewhat difficult problem, picturesquely described by the phrase "can you hear the shape of a drum?" arises when the solution of a partial differential equation (PDE) with specified boundary conditions is known and one is asked to find the shape and extent of the boundary. This may be termed the *boundary identification* problem. A special case of this problem is the *free boundary problem* and is characterized by the occurrence of frontiers or interfaces whose locations are *a priori* unknown. In the so-called *input identification* or *control* problem, one is asked to determine an input or control function that will yield a specified target solution to the problem. Another difficult problem, the *modeling problem*, arises when one is given noisy data observed over irregular intervals of space and time and is asked to develop a mathematical model to fit the observed data. With the advent of high-speed computers and artificial intelligence techniques, this modeling problem went through a metamorphosis and emerged as the *machine learning* problem.

This chapter touches upon a few problems in the evolution of these methods in the later half of the twentieth century starting first from a point of view propounded by Prof. Walter J. Karplus and finally looking at some of the current trends.

2. Nature of Inverse Problems

By their very nature, inverse problems are difficult to solve. Some times they are ill-posed. A *well-posed* mathematical problem must satisfy the following requirements: existence, uniqueness and stability. The existence problem is really a non-issue in many realistic situations because the physical reality must be a solution. However, due to noisy and/or insufficient measurement data, an accurate solution may not exist. More often, a major difficulty is to find a unique solution; this is especially so while solving a parameter identification problem. Different combinations of parameter values (including boundaries and boundary conditions) may lead to similar observations. One useful strategy to handle the non-uniqueness issue is to utilize *a priori* information as an additional constraint. These constraints generally involve the imposition of requirements such as smoothness on the unknown solution or its derivatives, or positivity, or maximum entropy or some other very general mathematical property. A more aggressive approach would be to use Bayesian approach and incorporate prior knowledge probabilistically. Indeed the well-known Tichonov regularization [Tichonov, '63] is a special form of Bayesian estimation theory. Another fruitful approach is via search. Given an observed data set, genetic algorithms and genetic programming can be used to probabilistically search a hypothesis space.

The quality of a solution to an inverse problem depends on the constraints imposed. The best constraints are those that not only seek good mathematical properties to the solution but also incorporate prior knowledge about the system. In the context of a problem in geophysics, for example, demanding non-negative permeability is an example of the former category and accommodating abrupt changes in the properties of rock formations (viz., discontinuities) would be an example of the later. It may be difficult indeed to accommodate both the smoothness constraints and the discontinuities simultaneously. That is to say, regularization may prevent the recovery of discontinuities. Recent advances in fractals may eventually provide a natural mechanism to incorporate the fractal or multi-scale nature of the structure of rocks and soils.

Inverse problems are often formulated by assuming that the underlying phenomenon is a dynamical system characterized by ordinary or partial differential equations, although no such assumption is always essential. In the context of remote sensing experiments, a mathematical formulation often leads to Fredholm integral equation of the first kind [Vemuri and Chen, '74; Vemuri and Jang, '91]. In both these formulations, often the goal is to build a *mathematical model* of the underlying phenomena. In some contexts a model is only a means to an end. Often, the ultimate goal in such cases is to test the validity of a hypothesis. In these cases, the model is used as a classifier (e. g., neural nets, decision trees) and it matters little whether the model is parametric or non-parametric; the classification accuracy becomes more important. From this point of view the entire field of *Machine Learning* [Mitchell, '97] can be treated as an exercise in solving inverse problems. *Data Mining*, a discipline aimed at finding hidden patterns, relations and trends also falls within the scope of inverse problems. While inverse problems associated with data mining represent data-rich situations, there is a class of inverse problems that are data-poor, such as the task of locating hidden structures in an enemy territory. Here the challenge is to combine general knowledge represented by models with specific knowledge represented by data.

3. Modeling an Aquifer

Inverse problems arising in geophysics are of particular interest in this chapter. Efficient environmental cleanup of subsurface chemical spills, enhanced oil recovery, safe containment of gases and fluids generated by underground nuclear tests, underground storage of nuclear waste, accurate characterization of water-supply aquifers - all require the modeling and simulation of the flow of fluids (air, water, contaminants) through porous media. The mathematical equations describing these processes are typically non-linear, non-homogeneous partial differential equations.

One of the first inverse problems I had the privilege of working is the task of estimating two parameters, namely transmissibility T (x, y) and storage coefficient S(x, y) of the unconfined aquifer in the San Fernando Valley groundwater basin (Figure 1). The starting point was the equation of water flow through porous medium:

$$\frac{\partial h}{\partial x}(T(x,y,h)\frac{\partial h}{\partial x})+\frac{\partial h}{\partial y}(T(x,y,h)\frac{\partial h}{\partial y})=S(x,y,h)\frac{\partial h}{\partial t}-Q(x,y,t) \quad \ldots(1)$$

Where $h(x, y, t)$ is the elevation of the water table above the mean sea level at the spatial point (x, y) at time t. The sign of the source term Q depends on the regimen used – pumping or recharge. Nonlinearities enter the equation because the transmissibility and storage coefficient, the two most important parameters, are mildly influenced by the elevation of the water table h, the dependent variable in the PDEs [Vemuri and Karplus, '69]. Irregularly sampled values of $h = h(x, y, t)$ were made available from historical well logs kept at the Department of Water and Power, City of Los Angeles. These wells were located at irregular points in the xy-plane and the well logs were recorded at irregular time intervals. The task had been to treat these data points as the sampled solution of the above equation.

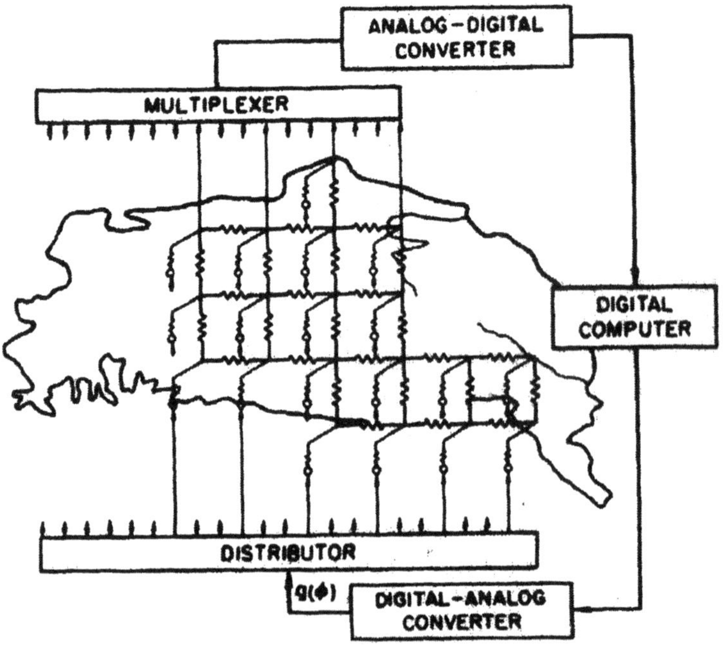

Figure 1. Hybrid Computer Modeling of San Fernando Valley Aquifer

What is left is to define the sources and boundary conditions. The aquifer was recharged periodically using recharge basins. During periods of surplus, water was brought in and allowed to flood large tracts of land, termed recharge basins, and the water was allowed to percolate. This information served to define a "distributed source." Fortunately, much of the San Fernando Valley is surrounded by hills permitting the assumption that the normal derivatives are zero along the impermeable rock formations. Admittedly, the quality and reliability of the data was poor and high-quality solutions were not expected.

All along, the goal had been to use this problem as a vehicle to test the suitability of a newly developed hybrid computer to solve partial differential equations. The hybrid computer, built in Dr. Karplus' laboratory by some of his graduate students, used a hardware subroutine (in this case, a resistance network, as depicted Figure 1) to invert matrices. The idea was to discretize the PDE in question using the usual finite differencing techniques but use the resistance network subroutine to solve the resulting system – rather than using the classical tridiagonal algorithm (in the one-dimensional case) or the alternating direction method (in higher-dimensional cases). As a resistance network relaxes almost instantaneously to its steady state, the solution of the linear tridiagonal or block-tridiagonal system can be obtained instantaneously – at least, theoretically!

With this computational tool in hand, what remained to be done was to solve the parameter identification problem as a multi-point boundary value problem on the hybrid computer. This was accomplished by solving Equation (1) forward in time for the value of h and its adjoint backward in time for the value of v. A gradient method was used to minimize an error defined in terms of the inner product $<h, v>$. As Equation (1) is non-linear the implementation required linearization and iteration.

Two major difficulties arose in the above process; one is theoretical and one practical. On the theoretical side, it was never clear where the physical boundaries of the aquifer were. The visible topographic boundary of the valley was clearly not the boundary of the aquifer. At this point, it was decided to include the boundary identification problem to be a part of the problem statement. An attempt was made to adjust the parameters and the boundary, one at a time using a heuristic approach interactively [Karplus and Vemuri, '67]. Although the results thus obtained were somewhat lackluster, the concept of using heuristics became a standard approach in AI and interactive inversion continues to be a topic of investigation to this day [Boschetti and Moresi, '02].

One of the causes for disappointment was advancing technology. Although the resistance network inverted the matrix in question instantaneously, the time it took for serial-to-parallel conversion of data, the mechanical crossbar multiplexer to step through the nodes of the resistance network and the subsequent re-conversion of the data from the parallel to serial format – not to mention the A/D and D/A conversion times - more or less sounded the death knell to the hybrid computer. As this work came to an end, solid-state multiplexers came into the market making the crossbar switch obsolete. The SDC 920 computer with its magnetic tape to store and read the programs already looked like a relic in the presence of the newly installed IBM 360 in a room down the hall.

4. Controlling Groundwater Contamination

This example is concerned with the remediation of groundwater in contaminated aquifers. The aquifer under study is a one square mile region once occupied by a petroleum depot in the 1940's. Due to the seepage of hydrocarbons from this facility, the aquifer, located some 90 to 180 feet beneath the ground surface was polluted. The primary goal was the containment of the pollutant.

Remediation is accomplished by pumping the contaminated water out, treating it and recharging the aquifer with the treated water. The problem is to determine the optimum placement of pumping (and recharge or injection) wells and optimum pumping (and recharge) schedules in order to achieve a set of objectives [Cedeño and Vemuri, '96].

Specifically, the problem solved was the determination of the optimum location of no more than 10 wells, on a 20 x 21 grid, so that three objectives are met. The first objective was to minimize the remediation cost that includes the capital cost for facilities, piping, and running costs associated with water treatment and day-to-day operations. Cost minimization was achieved by picking solutions that stay within a budget. The second objective was to maximize the amount of contaminant removed from the aquifer. This was straightforward and was obtained from the output of SUTRA code in kilograms. The third objective was to prevent unsafe levels of contaminant from leaving the site. The goal was to minimize the concentration of contaminant leaving the site as much as possible. This measure was also obtained from the output of SUTRA in parts per billion (ppb).

This problem was formulated as a multi-objective optimization problem and solved using a genetic algorithm (GA). The aquifer dynamics were simulated by repeatedly solving the PDEs describing the fluid flow using the U. S. Geological Survey's SUTRA code [Voss, '84]. SUTRA (Saturated, Unsaturated TRAnsport) is a 2-D hybrid finite-element/ finite-difference model aimed at solving the governing partial differential equations for groundwater flow and solute transport. As for boundary conditions, the northeast and southeast were treated as no-flow fault zones. Flux boundaries were assumed along the eastern and western edges of the site. A hydraulic conductivity of about 10 feet/day was assumed.

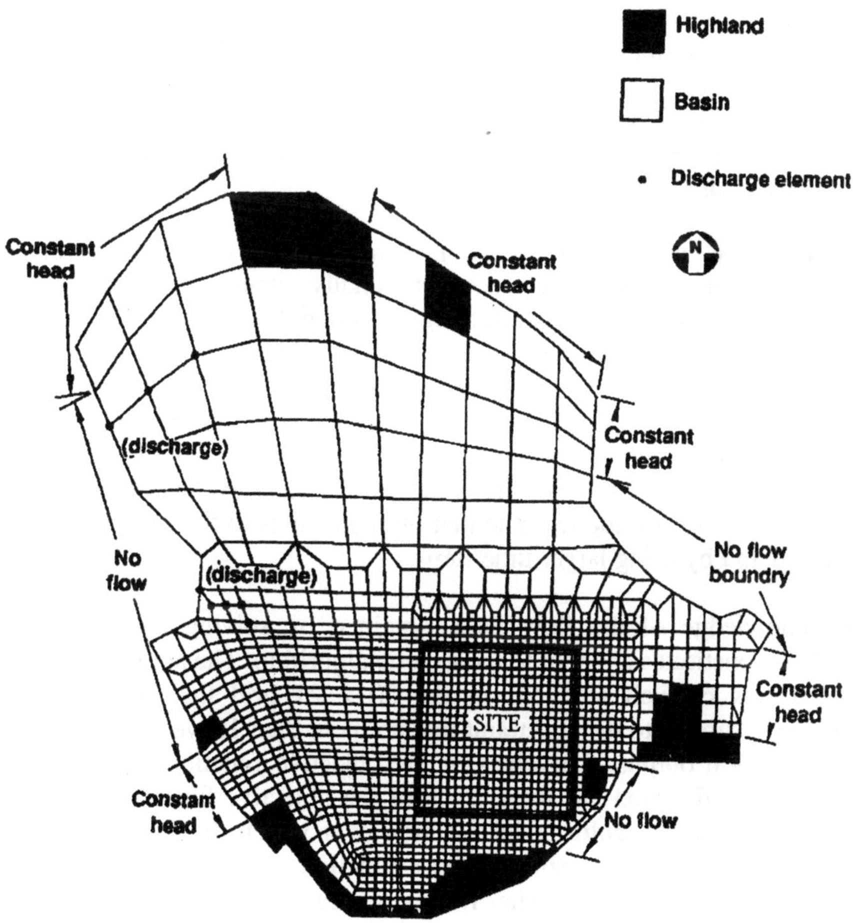

Figure 2. SUTRA nodes and elements in mesh

The simulation took into account the three "pump, treat and recharge" facilities already existing on the western edge of the site and the one facility on the eastern edge. Three more "pump, treat and recharge" facilities were planned. The task was to find an optimum location for these three new facilities. Based on data available, each facility was expected to cost $2.5 million with a capacity to clean 70 gallons of water per minute by reducing the concentration of contaminants from 550 ppb to negligible quantities. Thus the scenario described was as realistic as one can make it.

The computational mesh used had a total of 2436 nodes and 2385 elements, which covers the extent of the aquifer, which is much bigger than the one square mile region (see Figure 2). Out of this, a sub-region of size 20 x 21 nodes, covering an area of 5200 ft x 4950 ft. was chosen. Each node in this sub-region is considered as a potential pump location. The distance between nodes is 260 feet, and there are 420 possible pumping locations. Strictly speaking, the decision on where to locate the pumps should be governed by considerations such as the concentration of pollutants, pollution gradients, feasibility of drilling, and so on.

In this study another consideration, namely the computational burden, entered the picture. SUTRA is computationally intensive; it took about 6.5 minutes of elapsed time for evaluating the suitability of each proposed well configuration (called "fitness evaluation" in GA parlance). Each fitness evaluation ran SUTRA for 10 time steps, where each step represented one year. Any finer resolution meant much more time and adequate computational resources could not be mustered to do the job. It was felt that some gains in computational time in terms of fitness evaluations could have been made by restricting the potential pumping sites to the sub-region of the total grid and by using larger time steps.

This knowledge of the flow field, obtained by solving the PDEs using SUTRA, constitutes an input to the genetic algorithm. The algorithm's goal is to find sets of well locations in the 20 x 21 grid that best meets the objectives, subject to the constraint that no more than 10 wells were allowed. The algorithm outputs the values of the objective functions for a range of values of the decision variables. Each of the objective functions constitutes the "modes" of the multi-modal function over which the genetic algorithm conducts its search. No explicit fitness function, for use with the GA, was defined in this problem. The fitness of each individual solution was implicitly given by its objective values and its rank against other solutions.

Table 1: Sample solutions with three objectives

Gen.	No.of Wells	Contaminant leaving site (in ppb)	Contaminant Removed (in kg)	Cost Rank
10	7	6.30	36.71	6
10	6	7.79	39.62	5
18	7	11.15	43.37	4
19	7	13.27	59.02	3
20	7	2.44	64.68	1
21	7	1.21	81.98	2
23	9	6.59	98.89	7

Table 1 shows a set of solutions for a run of the GA. Only improvements from previous generations are shown. Only solutions meeting the regulatory limit and having a higher amount of contaminant removed from previous generations are shown. The last column ranks the solutions according to their cost. Lower rank values indicate lower cost solution. As expected, improvements in one objective had a negative impact in other objectives during the initial generations. The competition between solutions having above average ranking in particular objectives was visible again and as a result offspring with good ranking in must objectives appeared in later generations.

5. Iterative Inversion with Genetic Algorithms

An interesting problem in geophysics is to determine the properties of an object that cannot be directly observed. For example, one may be interested in obtaining information about the distribution of material under the ground, say the problem of locating underground structures. A related problem is that of locating hidden structures and facilities involved in clandestine operations. The fundamental difficulty in these problems is a lack of sufficient data spanning the entire spatial and temporal domains of interest. How does one understand a system that is too complex to sample or impossible to observe directly?

This problem becomes somewhat tractable if one has models capable of making good predictions under specific conditions. Then one can start with those baseline models and iteratively improve them by simultaneously incorporating new information as it becomes available. The new information can be obtained from a variety of sensors or by making, albeit inaccurate, predictions subjecting the available models to hypothetical stimuli.

It is relatively straightforward to gather field data from airborne surveys and use this data for inversion purposes. In the so-called *model-free* methods data from gravity or magnetic surveys can be used to find the distribution of material that satisfies the observations in a least-square sense. In the so-called *assumed source* methods, the unknown bodies are described with simple, regular shapes such as spheres, polyhedra, thin layers, etc., and the geometrical parameters of these objects and their locations are the target of inversion.

No matter what method is used, solution to a geophysical inversion problem is always non-unique. Additional subjective information is often necessary to resolve the problem further. Such additional information may be provided in the form of, say, (a) a specific starting model, (b) specific parameterization, restricting the search to predetermined geometric shapes, or (c) extra mathematical requirements by demanding solutions that exhibit some unique features (say, the set of all smooth solutions). However this leaves the actual extent of the *ambiguity domain* (i.e., the range of variability within the class of acceptable functions) unknown. Indeed, additional "knowledge" introduced into the problem in a quest to narrow down the scope of the inverse problem may introduce un-intended parameterization that may affect the final result. The question to be answered is: Is it possible that two radically different hypotheses (representing two completely different geological interpretations) satisfy the observed data exactly? In other words, how reliable is the inversion?

A possible solution to this dilemma is to conduct a systematic search within the ambiguity domain and guide this search with expert judgment. Such technique can be used to build a prototype and use the prototype model as a reference model during inversion. During ambiguity search, for instance, a misfit measure can be defined between the prototype and the inversion candidate. For example, the misfit could be the squared difference between the auto-correlation between the prototype model and the candidate inversion. Utility of a metric such as the above misfit measure may or may not be valid for the particular problem at hand. An alternative is to use global search using a method such as a genetic algorithm (GA).

In a GA, one starts with the solutions of a family of forward problems. Each solution is termed a chromosome. Problem parameters are genes and a concatenation of these parameters is the chromosome. Associated with each chromosome is a fitness function that describes how well that solution satisfies the requirements of the problem. Fitness functions can be defined in terms of subjective judgments and objective measures. Then those chromosomes that exhibit high fitness values are selected and a new family of solutions is generated through mathematical operations that are picturesquely termed mating, crossover and mutation.

6. Inverse Problems in Chemical Process Identification

A relevant problem of contemporary interest is the task of identifying what, if any, chemical warfare agents are being produced at some location. As many chemical processes are "dual-use" the same raw ingredients may go into the production of a fertilizer or an explosive; the difference is in the production pathways used by the processes in question. The goal here is to detect clandestine activities by observing and correlating multiple indicators or "signatures." As the best way to conceal an illegal activity is to make it look like a legal activity, it is not enough to look at obvious signatures such as components and their layout, traces left by testing activity, presence of precursors or degradation by-products in waste streams, and so on. It is also necessary to impose regularization constraints derived from prior subjective knowledge. Furthermore, each signature has its own precise pathway including the choice of technology. This is a difficult problem to solve and requires a careful formulation from first principles.

Success in this effort depends on acquiring a deeper understanding of the reaction pathways. A reaction pathway is nothing but a sequence of elementary reactions through which the precursors of a reaction (starting reactants) are routed through until the final target state is reached. The determination of these pathways entails two phases. The first phase entails the identification of all feasible candidate mechanisms and the second phase requires the selection of the ultimate pathway. Once an understanding is reached, mechanisms for more complex pathways can be determined through a synthesis of plausible elementary reactions. If a rigorous algorithmic method is available to perform this synthesis (which is known to be difficult due to its combinatorial complexity) then that gives an initial capability to explore.

Insight into this problem can be obtained by studying an analogous inverse problem associated with compartmental models that are very popular in biomedical engineering. In these models, some material enters the system from the external world (say food, medicine, etc.) and is transported through various compartments. There is no tangible output or observable other than the state of the system (e. g. the health) that can be deduced by observing by-products and effluents. Various nutrients, water, drugs and oxygen are typical inputs to the system and various excrements and carbon dioxide, urine, feces are typical effluents and energy is a typical output. Given some measurements on these inputs and outputs can one deduce information about the functioning of the body? This is what a physician does routinely? This task has become routine because of the accumulated knowledge about the anatomy and physiology of human body and the fact that the anatomy of a human body is fixed. This analogy stops here because there is some room for variability in the anatomical configuration of a chemical plant. Nevertheless, this would be a useful metaphor to gain an understanding of the problem.

References

[1] F. Boschetti and L. Moresi. "Interactive Inversion in Geosciences". *Geophysics,* 64:1226-1235, 2001.

[2] W. Cedeño. and V.R. Vemuri. "Genetic Algorithms in Aquifer Management". *Journal of Network and Computer Applications,* 19:171-187, 1996.

[3] W.J. Karplus. "Water Coning Before Breakthrough – an Electronic Analog Treatment". *Petroleum Trans. Amer. Soc. Mech. Engrs.*, 207: 240-245, 1956.

[4] W.J. Karplus and V. Vemuri. "Heuristic Optimization and Identification in Hybrid Field Simulation". *Proc. Fifth International Congress of AICA*, Lausanne, Switzerland, 345-350., 1967.

[5] T. Mitchell. *Machine Learning.* McGraw Hill, 1997.

[6] A.N. Tichonov. "Regularization of Incorrectly Posed Problems". *Soviet Math. Dokl.*, 4: 1624-1627, 1963.

[7] L. Tenorio. "Statistical Regularization of Inverse Problems". *SIAM Review,* 43(2): 347-366, 2001.

[8] V. Vemuri and W. J. Karplus. "Identification of Nonlinear Parameters of Ground Water Basins via Hybrid Computation". *Water Resources Research,* 5:172-185, 1969.

[9] V. Vemuri and F. P. Chen. "An Initial Value Method for Solving Fredholm Integral Equations of the First Kind". *Journal of Franklin Institute,* 297: 187-200, 1974.

[10] V. Vemuri. and G. S Jang. "Inversion of Fredholm Integral Equations of the first kind with fully connected Neural Networks". *Proc. SPIE Conference*, Orlando, FL. 1-5 April 1991.

[11] C.I. Voss. "SUTRA, Saturated-Unsaturated Transport: A finite-element simulation model for saturated-unsaturated, fluid-density-dependent groundwater flow with energy transport or chemically reactive single-species solute transport". *U. S. Geological Survey, Water Resources Investigations Report*, 84-4369, 1984.

10

Model Interoperability in the Discrete Event Paradigm: Representation of Continuous Models

Fernando J. Barros
Professor of Informatics Engineering
Universidade de Coimbra, Portugal
barros@dei.uc.pt

Bernard P. Zeigler
Professor of Electrical and Computer Enginnering
The University of Arizona, Tucson, AZ 85721
zeigler@ece.arizona.edu

Summary. Model interoperability requires that models can be mapped into a common formalism. To achieve this goal in the realm of discrete event systems we propose a standardization of the DEVS formalism. The DEVS formalism created in the earlier 70s has accommodated several changes since its creation. These changes have, for example, allowed expressing models with dynamic structure, and handling simultaneous events. We overview different DEVS flavors and we present both basic and port/link based formalisms. To increase the scope of model interoperability we also present guidelines for transforming continuous models specified in the Continuous Flow System Specification formalism into discrete event models. This adds to existing alternatives for simulating continuous systems with discrete event approaches whose asynchronous time flow supports efficient parallel and distributed execution. Establishing the equivalence relationships between various formalisms, both existing and new, within the DEVS formalism will help to standardize the field of discrete event simulation and eventually will lead to a high degree of compatibility among models, with a strong impact in the reuse of model libraries

1. Introduction

Interoperability at the modeling level requires a common formalism in which different type of models can be interconnected and interoperated. Interoperability at this level can be distinguished from interoperability at the simulation level which is the primary concern of current interoperability efforts surrounding the High Level Architecture standard [11]. More specifically, interoperability at the simulation level is concerned with interoperating simulators that are described by interfaces encapsulating and hiding their underlying models. In contrast, interoperability at the modeling level is concerned more fundamentally with integrating models that are usually developed in different paradigms and expressed in different formalisms, such as analog/differential equations or digital/state machines. For more discussion on the different levels of interoperability see [18].

The Discrete Event System Specification (DEVS) formalism was developed in the middle 70's by Zeigler [20]. DEVS was primarily created to describe the fundamental aspects that characterize discrete event systems. Since then DEVS has been revised to express and exploit simultaneous events [10], dynamic structure models [1], and port based communication [2], [17], [18], [19], [22] to name a few changes. Mapping of other discrete event formalisms have also been described in [14]. The use of DEVS to represent continuous systems has also been achieved [12], [15], [22].

In this article we present a standardization proposal of the DEVS formalism based on the Discrete Flow Systems Specification formalism (DFSS). The DFSS formalism can represent dynamic structure models, and a class of models, called mobile components [4]. Ports and links are widely used in modular frameworks for discrete event simulation. A definition of a standard should also include these elements. We present also a port/link version of the DFSS formalism.

The Continuous Flow System Specification [5] is a basic framework for expressing numerical hierarchical and modular multirate integration methods. To increase the scope of model interoperability we present guidelines to convert hierarchical and modular CFSS networks into DEVS/DFSS networks. The goal is to integrate continuous and discrete models into the same framework thus increasing the family of models that can interoperate in the same basic DEVS/DFSS framework.

2. DEVS Atomic Models

There are several definitions for DEVS atomic models. We shall show that these variants do not cause difficulty in model interoperability because they all express models that are equivalent to the following proposed DEVS standard.

2.1 Standard DEVS

The proposed DEVS standard includes the specification of model initial state and the use of a single transition to replace both internal and external transition functions. These changes have been introduced in DFSS and allow the specification of mobile components [4]. In the standard DEVS formalism, atomic models are formally defined by:

$$M = (X,Y,S,q_0,\tau,\delta,\lambda)$$

where

X is the set of input values

Y is the set of output values

S is the set of partial states (p-states)

$Q = \{(s,e) | s \in S, 0 \leq e \leq \tau(s)\}$ is the state set

$q_0 = (s_0,e_0) \in Q$, is the initial state, with

s_0 the initial p-state (partial state)

e_0 the time elapsed in the initial p-state s_0

$\tau: S \rightarrow \mathbf{R}_0^+$ is the time advance function

$\delta: Q \times X^\phi \rightarrow S$ is the transition function, where

$X^\phi = X \cup \{\phi\}$

ϕ is the null event (marker for absence)

$\lambda: S \rightarrow Y$ is the partial output function

The output function, $\Lambda: Q \rightarrow Y^\phi$, is defined by

$$\Lambda(s,e) = \begin{cases} \lambda(s) & \text{if } e = \tau(s) \\ \phi & \text{if } e < \tau(s) \end{cases}$$

The semantics of the formalism are briefly described as follows: If no event arrives to the system, it will stay in partial state s for time $\tau(s)$ during which the elapsed time, e will accumulate. When $e = \tau(s)$, the system changes to the state $(\delta((s,\tau(s)),\phi),0)$, an internal event. If an external event, $x \in X$, arrives when the system is in the state (s,e) with $e \leq \tau(s)$ it will change to the state $(\delta((s,e),x),0)$. Included in this specification, is the "confluent" situation where an external event, $x \in X$, arrives at the same time as an internal event is about to take place (when $e = \tau(s)$). In this case, the system changes to the state $(\delta((s,\tau(s)),x),0)$ which the modeler can specify in a distinct manner if appropriate. We note that the transition function is used to determine all state changes and it is defined for all allowed combinations of state and input conditions.

The Appendix provides examples of models expressed in the formalisms discussed in this article. For a more complete introduction to discrete event formalisms and theory see [22].

2.2 Parallel DEVS

To facilitate modeler convenience, some existing variants of the DEVS formalism separate out the distinct transition situations. The parallel DEVS [10] has been developed to handle model transitions occurring when external input values arrive when time to next transition has elapsed. To achieve this goal Parallel DEVS (P-DEVS) provides a particular structure to the transition function that is specified by three functions: the internal transition δ_{int}, the external transition δ_{ext}, and the confluent transition δ_{con}. The P-DEVS is thus defined by:

$$P = (X,Y,S,q_0,\tau,\delta_{int},\delta_{ext},\delta_{con},\lambda)$$

where

δ_{int}: $S \rightarrow S$, is the internal transition function
δ_{ext}: $Q \times X \rightarrow S$, is the external transition function
δ_{con}: $S \times X \rightarrow S$, is the confluent transition function

From the P-DEVS = $(X,Y,S,q_0,\tau,\delta_{int},\delta_{ext},\delta_{con},\lambda)$, we can derive the basic model $M = (X,Y,S,q_0,\tau,\delta,\lambda)$

where

$$\delta((s,e),x) = \begin{array}{ll} \delta_{int}(s) & \text{if } x = \phi \text{ and } e = \tau(s) \\ \delta_{ext}((s,e),x) & \text{if } x \neq \phi \text{ and } e \neq \tau(s) \\ \delta_{con}(s,x) & \text{if } x \neq \phi \text{ and } e = \tau(s) \end{array}$$

Although such separation helps in the model building exercise, for a unified standard, we believe that the single integrated form is best since it make it easier to develop more advanced aspects of the standard.

2.3 Structured DEVS

The *Structured Discrete Event System Specification* (SDEVS) provides a fine-grained control over models. The transition function, for example, can be decomposed into a set of transitions, one for each p-state. The same applies to the other functions. In the SDEVS formalism [6], a model as described by:

$$\mathbf{S} = (X,Y,S,q_0,\kappa,I,\{\tau_i \mid i \in I\},\{\delta_i \mid i \in I\},\{\lambda_i \mid i \in I\})$$

where

I is a set of indexes
κ: $S \rightarrow I$ is the index function

A SDEVS model $\mathbf{S} = (X,Y,S,q_0,\kappa,I,\{\tau_i\},\{\lambda_i\},\{\delta_i\})$, is equivalent to the model $M = (X,Y,S,q_0,\tau,\delta,\lambda)$, where for all $s \in S$, $\tau(s) = \tau_{\kappa(s)}(s)$, $\delta((s,e),x) =$

$\delta_{\kappa(s)}((s,e),x)$ and $\lambda(s) = \lambda_{\kappa(s)}(s)$, [7].

2.4 *Port-based models*

Ports are used in most modular discrete event modeling simulation environments. They provide structure to input and output values and they enable a convenient graphical representation of models similar to the usual one provided for hardware components. Formally a port-based DEVS can be described by

$$M = (X,Y,S,q_0,\tau,\delta,\lambda)$$

where

$X = \{(p,v) | p \in PX, v \in \text{bags-of-set}(VX_p)\}$
$Y = \{(p,v) | p \in PY, v \in \text{bags-of-set}(VY_p)\} \cup \{\phi\}$

and

PX is the set of input ports
VX_p is the set of values of input port p
PY is the set of output ports
VY_p is the set of values of output port p

The notation bags-of-set(A) refers to the set of all bags made of elements of A, excluding the empty bag.

All the models presented in the last sections can easily be modified to accommodate a port-based definition of input and output values.

Although port based models can be represented uniquely as basic models, the converse does not hold, for it not possible to uniquely impose a port-based structure on a non-structured interface. Thus an important decision for establishing a standard is to decide whether models should have a port-based interface. This decision has a great impact on model definition because, in general, models that are compatible in the abstract, may not be compatible under different choices of ports and values.

3. The DFSS Network Model

The DFSS formalism provides a representation of dynamic structure discrete event models [6]. DFSS network models exhibit a time-varying structure. Formally, a DFSS network is defined by:

$$DFSS_N = (X_N,Y_N,\eta,M_\eta)$$

where

N is the network name
X_N is the set of network input values
Y_N is the set of network output values
η is the name of the dynamic network executive
M_η is the model of the executive η

The model of the executive, is a modified basic model defined by

$$M_\eta = (X_\eta, Y_\eta, S_\eta, q_{0,\eta}, \gamma, \Sigma^*, \tau_\eta, \delta_\eta, \lambda_\eta)$$

where

$\gamma: S_\eta \rightarrow \Sigma^*$ is the structure function
Σ^* is the set of network structures

A structure $\Sigma_j \in \Sigma^*$ associated with the executive partial state $s_{j,\eta} \in S_\eta$, is given by

$$\Sigma_j = \gamma(s_{j,\eta}) = (D_j, \{M_{i,j}\}, \{I_{i,j}\}, \{Z_{i,j}\})$$

where

D_j is the set of component names associated with the executive partial state $s_{j,\eta}$
for all $i \in D_j$, $M_{i,j}$ is the model of component i
for all $i \in D_j \cup \{\eta, N\}$, $I_{i,j}$ is set of components influencers of i
for all $i \in D_j \cup \{\eta\}$, $Z_{i,j}$ is the input function of component i
$Z_{N,j}$ is the output function of the network N

These variables are subject to the following constraints for every $s_{j,\eta} \in S_\eta$:

$\eta, N \notin D_j$
$N \notin I_{N,j}$
$M_{i,j} = (X_{i,j}, Y_{i,j}, S_i, q_{0,i}, \tau_i, \delta_{i,j}, \lambda_{i,j})$ is a basic model, for all $i \in D_j$, with
$\delta_{i,j}: Q_i \times X^\phi_{i,j} \rightarrow S_i$

For notational simplicity we constrain the partial state set, the initial state and the time to output function to remain constant.

$$Z_{i,j}: \bigtimes_{k \in I_{i,j}} V_{k,j} \rightarrow X_{i,j}, \text{ for all } i \in D_j \cup \{\eta\}$$

where

$$V_{k,j} = \begin{cases} Y_{k,j} & \text{if } k \neq N \\ X_N & \text{if } k = N \end{cases}$$

The Network output function is given by

$$Z_{N,j}: \bigtimes_{k \in I_{N,j}} Y_{k,j} \rightarrow Y_N$$

The components input function and network output function are subjected to the following constraint:

$Z_{i,j}(\phi, \ldots, \phi) = \phi$, for all $i \in D_j \cup \{\eta, N\}$

To simplify the notation, in the remaining part of the article we consider only networks with a static structure, i.e., we consider that $\Sigma = \gamma(s_\eta) = (D,\{M_i\},\{I_i\},\{Z_i\})$ for all $s_\eta \in S_\eta$. The extension of the discussion to dynamic structure models is straightforward.

4. DFSS Port-Based Networks

Network of port-based models are commonly described using the concept of link that represents a flow of information between two ports. A link-based definition the DFSS network is given by:

$$\gamma(s_\eta) = (D,\{M_i\},L)$$

where

L is the set of network links

Links are used by almost every modular discrete event modeling framework. Usually, links provide a shorter notation for an identity function that keeps values flowing through links unchanged. Formally a link is an ordered pair relating two ports. A link l connecting port i of component A to port j of component B is noted by $l = ((A,i),(B,j))$ and is graphically represented in Figure 1.

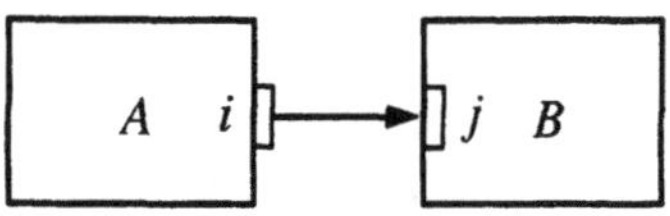

Figure 1. Link from *A* to *B*

Link based definitions require that component models, either atomic or network, are themselves expressed in port based notation.

4.1 From Links to DFSS Network Models

Link-based networks can be expressed in the basic DFSS formalism. The transformation involves the definition of the sets $\{I_i\}$ and $\{Z_i\}$ from the set L of links. We will describe this mapping by an example rather than by a more formal and general definition. The example we will use is depicted in Figure 2 that represents a network with four components where component D is connected from components A, B and C.

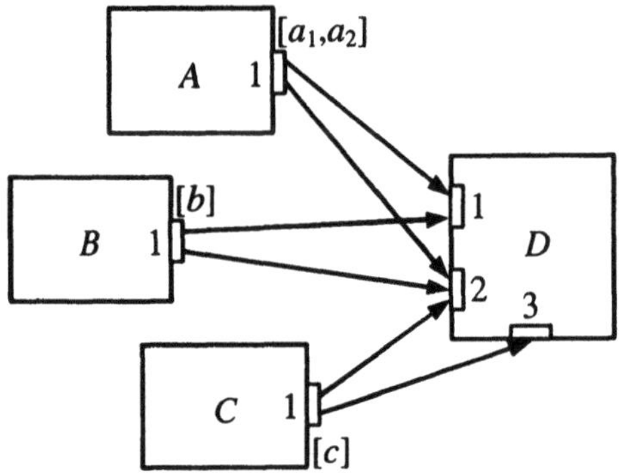

Figure 2. Network defined by ports and links

The set L of network links is given by

$$L = \{((A,1),(D,1)),((A,1),(D,2)),$$
$$((B,1),(D,1)),((B,1),(D,2)),$$
$$((C,1),(D,2)),((C,1),(D,3))\}$$

A family of link-based formalisms can be derived from DFSS networks. These formalisms will essentially differ in the definition of the input function associated with the link set.

We derive now the DFSS standard network definition from the link set. The influencees of a component X can be derived from the set of links L as the set of components that have at least one link to component X. In our example the influencees of component D are given by

$$I_D = \{A,B,C\}$$

This set allows depicting the network using the basic DFSS notation to networks represented in Figure 3.

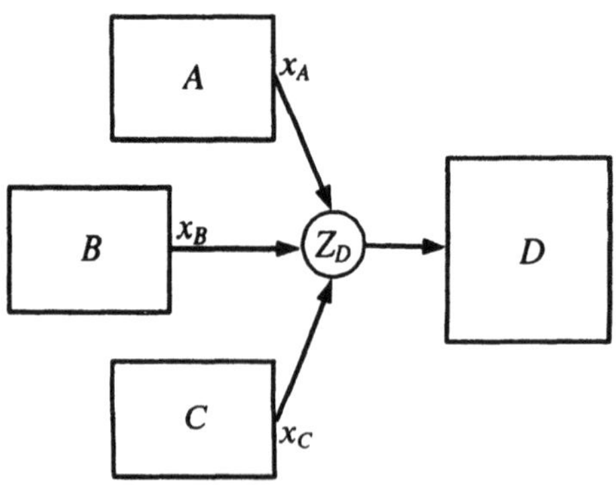

Figure 3. DFSS network representation

Built upon link information, set $\{Z_i\}$ is open to numerous instantiations. An useful choice, when dealing with port-based models, is a set of pairs given by (*port*, *bag-of-values*), where values on the same port are grouped in the same bag. In the network of Figure 2, the values on port 1 of components *A*, *B*, and *C* are $[a_1,a_2]$, $[b]$ and $[c]$, respectively. The input function Z_D is given by

$$Z_D(x_A,x_B,x_C) = Z_D((1,[a_1,a_2]),(1,[b]),(1,[c])) = \\ \{(1,[a_1,a_2,b]),(2,[a_1,a_2,b,c]),(3,[c])\}$$

The input function must obviously obey the constraints stated in Section 3. This function has special cases when any of the input values is null (ϕ). When all input values are null the function evaluates to null. In our example if $a_1 = a_2 = b = c = \phi$, the input function is given by

$$Z_D(x_A,x_B,x_C) = \phi$$

This function also removes all null values. For example if $a_1 = a_2 = \phi$ the function evaluates to

$$Z_D(x_A,x_B,x_C) = \{(1,[b]),(2,[b,c]),(3,[c])\}$$

The input function also removes ports with only null values. For example, if $c = \phi$, the input function removes port 3, that receives only null values. The function evaluates thus to

$$Z_D(x_A,x_B,x_C) = \{(1,[a_1,a_2,b]),(2,[a_1,a_2,b])\}$$

The bag definition permits to keep copies of the same input value. When $a_1 = a_2 = b = c = x$ the input function evaluates to

$$Z_D(x_A,x_B,x_C) = \{(1,[x,x,x]),(2,[x,x,x,x]),(3,[x])\}$$

As indicated, the function just described is just one possible definition of the input function. However, for achieving compatibility among models, a standardization of this function needs to be achieved.

5. The CFSS Formalism

The Continuous Flow System Specification formalism (CFSS) provides a representation of continuous systems based on a generalized version of the sampling concept. Formally, a *Continuous Flow System Specification* is given by [3]:

$$CFSS = (X,Y,S,q_0,\rho,\delta,\Lambda)$$

where

X is the set of input values

Y is the set of output values
S is the set of partial states (p-states)
$Q = \{(s,e) | s \in S, 0 \leq e \leq \rho(s)\}$ is the state set
$q_0 = (s_0,e_0) \in Q$, is the initial state, with
s_0 the initial p-state
e_0 the time elapsed in p-state s_0
$\rho: S \rightarrow \mathbf{R}_0^+$ is the time to input function
$\delta: S \times X \rightarrow S$ is the transition function
$\Lambda: Q \rightarrow Y$ is the output function

A CFSS component samples its input accordingly to the time to input function ρ that represents the time to the next sample. Figure 4 depicts the output trajectory of a CFSS component.

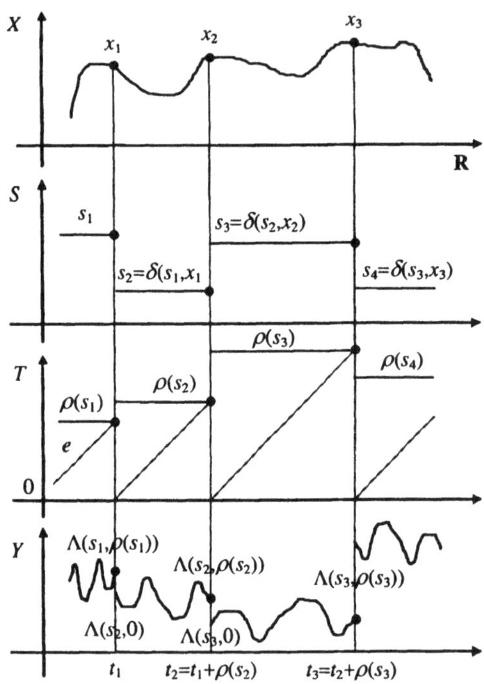

Figure 4. CFSS trajectories.

At time t_1 the input flow is sampled and the model changes from state s_1 to state $s_2 = \delta(s_1,x_1)$, where x_1 is the value of the input flow at time t_1. We note that by Assumption 2, s_1 is the p-state at time t_1, and the output at this time is given by $\Lambda(s_1,\rho(s_1))$. The time to read the next input is given at p-state s_2 by $\rho(s_2)$. After this time interval the input is sampled again and a new transition occurs at time $t_2 = t_1 + \rho(s_2)$. At time t_2 the output is given by $\Lambda(s_2,\rho(s_2))$ and immediately after t_2, at time t_2^+, the output is given by $\Lambda(s_3,0)$, where $s_3 = \delta(s_2,x_2)$, and x_2 is the input value at time t_2. The output trajectory is a function of both the p-state and the time elapsed in the current p-state. We note that, although the p-state is piecewise constant, the output can be piecewise continuous since it is a function of the elapsed time which is a real value. Of course the output is only manifest as at discrete instants when other components sample it.

5.1 CFSS Networks

Formally, a Continuous Flow System Specification Network is a defined by the 4-tuple [5]:

$$CFN_N = (X_N, Y_N, \eta, M_\eta)$$

where

N is the network name
X_N is the set of continuous input flow values
Y_N is the set of continuous output flow values
η is the executive name
M_η is the model of the executive η

The model of the executive is a modified CFSS, defined by

$$M_\eta = (X_\eta, Y_\eta, S_\eta, q_{0,\eta}, \Sigma^*, \gamma, \rho_\eta, \delta_\eta, \Lambda_\eta)$$

where

Σ^* is the set of network structures
$\gamma: Q_\eta \rightarrow \Sigma^*$ is the structure function

The network structure $\Sigma_{j,e} \in \Sigma^*$, corresponding to the state $(s_{j,\eta}, e) \in Q_\eta$, is given by the 4-tuple:

$$\Sigma_{j,e} = \gamma(s_{j,\eta}, e) = (D_j, \{M_{i,j,e}\}, \{I_{i,j}\}, \{Z_{i,j,e}\})$$

where

D_j is the set of component names associated with the executive partial state $s_{\eta j}$

for all $i \in D_j$
$M_{i,j,e}$ is the model of component i
$I_{i,j}$ is the set of components, the ***influencers*** of i
$Z_{i,j,e}$ is the input function of component i

For simplicity we consider that models can only change with the change of the partial state, i.e., $M_{i,j,e} = M_{i,j}$.

Variables are subjected to the following constraints for every $q_{j,\eta} \in Q_\eta$:

$\eta \notin D_j$
$N \notin I_{N,j}$
$M_{i,j} = (X_{i,j}, Y_{i,j}, S_i, q_{0,i}, \rho_i, \delta_{i,j}, \Lambda_{i,j})$ is a CFSS model, for all $i \in D_j$, with
$\delta_{i,j}: Q_i \times X_{i,j} \rightarrow S_i$

For notational simplicity we constrain the partial state set, the initial state and the time-to-output function to remain constant.

$$Z_{N,j,e}: \underset{k \in I_{N,j}}{\times} Y_{k,j} \rightarrow Y_N$$
$$Z_{i,j,e}: \underset{k \in I_{ij}}{\times} V_{k,j} \rightarrow X_{i,j}, \text{ for all } i \in D_j \cup \{\eta\}$$

where

$$V_{k,j} = \begin{cases} Y_{k,j} & \text{if } k \neq N \\ X_N & \text{if } k = N \end{cases}$$

5.2 *Mapping CFSS Networks into DEVS*

The representation of hybrid systems has been the subject of many studies and several solutions were proposed ([5], [12], [13],[16], [21]). Recent research within the DEVS framework has developed representations of continuous models by discrete event models. One solution is termed *quantization* in which continuous signals are quantized into piecewise constant segments ([15]). Discrete event values are thus sent when a change of the segment value occurs.

The Continuous Flow System Specification (CFSS) was introduced in [3] as a formalism to represent continuous systems based on a generalized version of the sampling concept. The formalism supports variable rate sampling and multi-sampling. These characteristics permit the development of variable-step numerical integration methods. The combination of CFSS and DEVS permits also the representation of hybrid systems [6] that can combine numerical integration methods with digital controllers.

To promote model interoperability we exploit here the ability to represent CFSS components within the DEVS formalism. The transformation of CFSS networks into DEVS networks involves both connections and components. The general idea is to transform each connection in the original CFSS network in two connections in the equivalent DEVS network, one connection for requesting the value and the other to send the requested value. Also each component in the original network needs to be transformed into a DEVS model to be able to handle discrete inputs. To illustrate these steps we use the coupled model of Figure 5, where for simplicity the network executive is omitted since this discussion is restricted to static structure models. In this network component B samples component A. that is treated as a continuous signal generator.

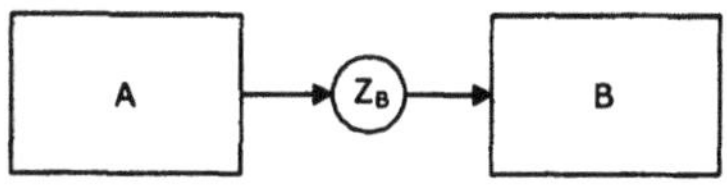

Figure 5. CFSS network (I)

An equivalent DEVS network is depicted in Figure 6, where the original CFSS components A and B were replaced by the DEVS components A' and B' and the original function Z_A was replaced by functions $Z_{A'}$ and $Z_{B'}$. For simplicity we assume that input functions are the identity function.

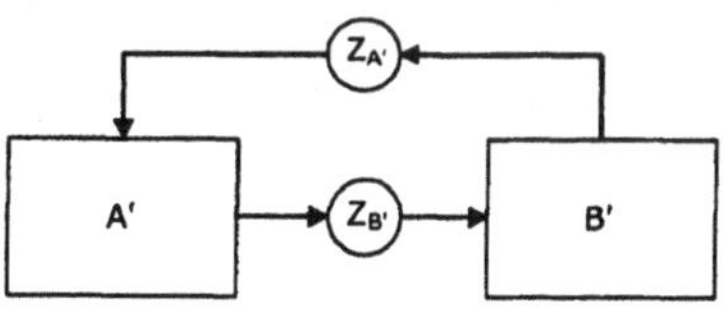

Figure 6. Equivalent DEVS network

The behavior of CFSS component B is illustrated in Figure 7. At time t, $e = \rho(s_1)$ and component changes from p-state s_B to p-state $s' = \delta(s_B, x)$, where x is the value sampled from component A at time t.

For obtaining a DEVS network with the same behavior, whenever CFSS component B samples its input, the DEVS corresponding component B' must send a discrete event output to component A' requesting a value. When component A' receives this request it sends a DEVS output with the required value x, the value that was sampled in the original CFSS network.

The state diagrams of DEVS components A' and B' are depicted in Figure 8, where $t^{n+} = t + n.\Delta t$, $n \geq 0$, $\Delta t > 0$, $\lim \Delta t \rightarrow 0$.

DEVS component B' sends its request at time t where $e = \tau(s_I)$. It then passivates in a state s_{II}, i.e., $\tau(s_I) = \infty$. Component A' goes into a transitory state and produces value x and then it passivates again.

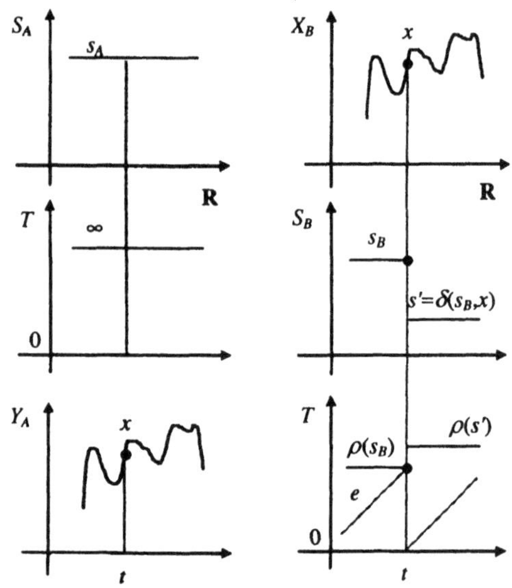

Figure 7. Behavior of the CFSS component B

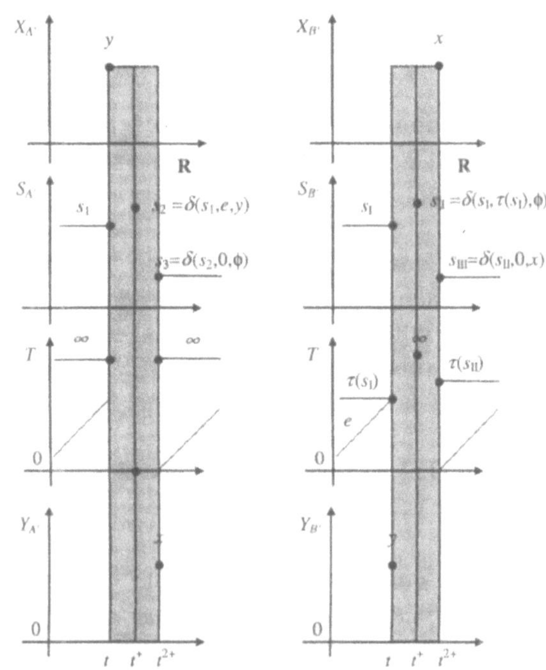

Figure 8. Behavior of the DEVS components A' and B'

A more complex situation arises when a component is being sampled by two or more components, as in the network of Figure 9. In the equivalent DEVS network component A needs now to take care where signals were originated and to send only values to the requesting component. This can be done in a number of ways. For example, B and C can send sample requests to distinct input ports of A (e.g., fromB and fromC) which can place outputs on corresponding output ports (e.g., toB and toC). The latter can be coupled to corresponding input ports of B and C, respectively. We note that various implementations of request/reply protocols correspond to different management services for distributing data among federates in distributed simulations, such as publish/subscribe in HLA [11] and the event-channel in Real-time CORBA [18].

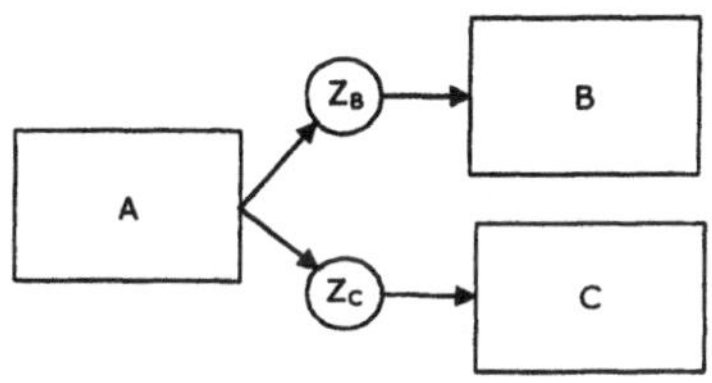

Figure 9. CFSS network (II)

An example of a multirate system can be found in Appendix.

6. Conclusions

6.1 Representation of Continuous Systems

As pointed out before, the CFSS formalism can represent numerical methods with variable steps. These methods can largely outperform conventional simple step methods. Traditional approaches of variable step methods do not fully exploit these capabilities. Standard variable step methods use the same integration step for all models; although the time step is allowed to vary from step to step. The CFSS formalism allows each model to have its own and independent integration step. Thus the formalism can exploit an efficient integration step for all the differential equations describing a system [3]. Thus the CFSS formalism offers an alternative to the use of quantized integration. The latter also can be viewed as a species of variable time step method whose time steps are determined by predictions of quantum threshold crossing rather than local error control. Both share the advantages of freeing individual integrators from the constraints of lock-step synchronization. And both employ the elapsed time variable to compute accurate updates when needed (quantization does this for state due to input events; CFSS does this when sampling from an output trajectory).

Automatic transformation of CFSS models into DEVS models will offer modelers more capability to express continuous and hybrid models in familiar terms while gaining the advantages of efficient discrete event simulation. Among these advantages is the potential to exploit the performance of current parallel and distributed discrete simulation environments (see [22] for example).

6.2 *Model interoperability*

The basic DFSS/DEVS framework can represent families of both atomic and coupled DEVS formalism. Structured DEVS, parallel DEVS and DEVS-with-ports can be represented by the basic DEVS formalism. Establishing the equivalence between new formalisms and the basic DEVS formalism will help to standardize the field of discrete event simulation and eventually will lead to a high degree of compatibility among models, with a strong impact in the reuse of model libraries. We have outlined the necessary transformations to convert CFSS models into DEVS models. These transformations allow numerical multirate integration methods to be integrated within the DEVS based frameworks. These transformations will allow CFSS models to interoperate with quantized integrators and other DEVS models.

Acknowledgements

The work of the first author was partially funded by the Portuguese Science and Technology Foundation under project PRAXIS/EEI/14152/98. The research of the second author has been supported in part by NSF Next Generation Software (NGS) "A Simulation Platform for Experiments and Evaluation of Distributed-Computing Systems (SPEED-CS)" Grant No. EIA-9975050, and NSF Scalable Enterprise Systems (SES) grant, "Discrete Event System Specification (DEVS) as a Formal Modeling and Simulation Framework for Scaleable Enterprise Design," Grant No. DMI-0122227.

Appendix: Model Examples

A.1 Atomic Model: Fire-Once Neuron

In a DEVS, inputs events arriving in time are handled somewhat in the manner of interrupts by the modeler-specified external transition function and result in an immediate change in state. This function determines the state transition and how long to stay in the new state. At the end of this duration, the model outputs an event determined by the output function and transits to a new state determined by the internal transition function. As an example, consider a DEVS model of the *fire-once neuron* [Zeigler, 2002]. It remains in the *receptive* state until an input arrives. The external transition function sends it to the *fire* state, where it remains for a duration given by firing-time, after which it emits a pulse and enters state *refract*. The time advance in *refract* is infinity, and it remains in *refract* upon receiving an input, so the neuron remains in *refract* forever, never able to fire again.

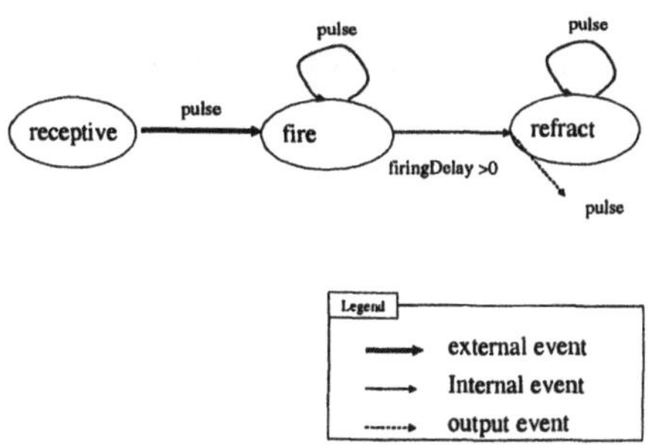

Figure 10. Fire-Once Neuron

Figure 10 illustrates the use of graphical notation to denote critical aspects of DEVS models. We'll correlate this notation with the DEVS model for the fire-once neuron, which can be expressed as follows:

$$M = (X, Y, S, \delta_{int}, \delta_{ext}, \lambda, ta)$$

where

X is the set of input values, {pulse}

Y is the set of output values, {pulse}

S is a set of states, {receptive, fire, refract}

δ_{int}: $S \rightarrow S$ is the internal transition function

δ_{int}(fire) = refract

(note: internal transitions are depicted by light transition arrows as in Figure 10)

δ_{ext}: $Q \times X^b \rightarrow$ S is the external transition function,

δ_{ext}(receptive, e, pulse^b) = fire

(note: external transitions are depicted by heavy transition arrows as in Figure 10)

δ_{ext}(fire, e, pulse^b) = fire
δ_{ext}(refract, e, pulse^b) = refract

(note: we use the notation, pulse^b, to indicate one or more pulses arriving at the same instant. In this model, they have no combined effect greater than any one of them.)

$\lambda: S \rightarrow Y^b$ is the output function
λ(fire) = {pulse}

(note: outputs are depicted by dotted arrows as in Figure 10; also only a single pulse is generated.)

λ(receptive) = φ (the symbol for null output, i.e., no event occurs)
λ(refract) = φ

ta: $S \rightarrow \mathbf{R}^{+}_{0,\infty}$ is the *time advance* function

ta(receptive) = ∞
ta(fire) = firingDelay
ta(refract) = ∞

Note that states that have an infinite time advance cannot generate output (since an internal transition will never occur from such a state). So we need not explicitly associate the null-event output with such states (as we have done for illustration above). In other words, the output function need only be defined for the non-passive states of a model.

A.2 Coupled Model: Neural Network

Neural networks can be represented as coupled models within the DEVS formalism. Moreover, this possibility gives rise to a whole class of networks constructed of discrete event neuron models that includes the traditional models as a subclass. More significantly, this class includes new kinds with distinct behaviors and information processing properties. Figure 11 illustrates a small network of fire-once neurons (as in Figure 10). This kind of network can compute the shortest path in a directed graph by mapping distances of edges to an equivalent time value. This time is then assigned as the fireDelay for a neuron which represents the edge. In Figure 11, for example, a pulse emitted from the generator explores two paths concurrently to reach the final neuron (number 4). Depending of the summed firing delays along each path, a pulse emerging from one or the other will arrive earlier to the final neuron, representing the shortest path of an associated digraph. The path can be reconstructed by extending the neuron model to allow retracing the path of earliest firings. If instead of shortest paths, we request longest paths, a net of neurons that fire after their assigned fireDelays *every* time they receive a pulse will do the job. Interestingly, finding critical paths in PERT charts require such longest path computation.

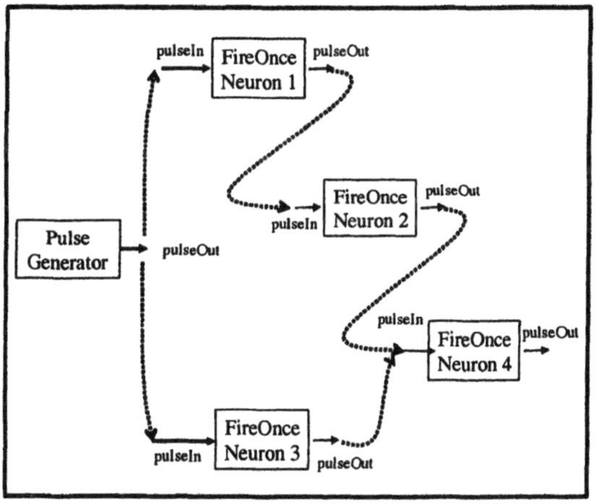

Figure 11. Neural Network Coupled Model

A.3 DEVS Representation of ODEs

Figure 12 illustrates a general mapping of ordinary differential equation (ODE) systems to DEVS representations. The instantaneous functions in a ODE system are mapped to discrete event equivalents that retain input information from event to event. The integrators of the ODE are mapped to DEVS equivalents, called *quantized* integrators. The couplings among instantaneous functions and integrators in the ODE systems are preserved in the DEVS mapping. Since this is a structure-preserving mapping, it does not require apriori solution of the ODE to make it work, and thus offers a practical and efficient way to simulate ODE systems.

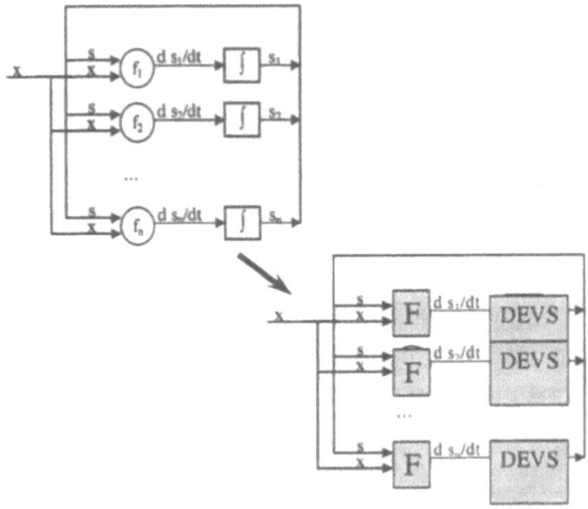

Figure 12. Mapping Differential Equation Systems to DEVS Simulators

The DEVS realization of the quantized integrator has the simple definition:
$M = (X,Y,S,\delta_{int},\delta_{ext},\delta_{con},\lambda,ta)$
where

$X = Y = R$ and $S = R \times R \times I$ and

- $\delta_{int}(q,x,n) = (n+D*\text{sign}(x),x,n+\text{sign}(x))$
- $\delta_{ext}((q,x,n),e,x') = (q+x*e,x',n)$
- $\delta_{con}((q,x,n),x') = (n+D*\text{sign}(x), x',n+\text{sign}(x))$
- $\lambda(q,x) = (n+\text{sign}(x))*D$
- $ta(q,x,n) = |((n+1)*D-q)/x|$ if $x > 0$ and $(n+1)*D-q > 0$
  $= |(q-n*D)/x|$ if $x < 0$ and $|q-n*D| > 0$
  $= |D/x|$ if $x \neq 0$ and none of the above
  $= \infty$ otherwise (i.e., $x = 0$)

Here we keep track of the boundary below (or at) the current state q, i.e., the integer floor(q/D). Recall that even if we start on a boundary, the state may eventually be inside a block (hence not a multiple of D) as a result of an external transition. If, as in Figure 13a), we are on a boundary, the time advance computation merely divides D by the current input x (which is the derivative, or slope, after all). If we reach the upper boundary (n+1) or lower boundary (n −1) we output and update the state accordingly. Note that so long as the input remains the same, the time to cross successive boundaries will be the same. Figure 13b) shows that when a new input is received, we update the state using the old input and the elapsed time. From this new state, q, the new time to reach either the upper or lower boundary is computed.

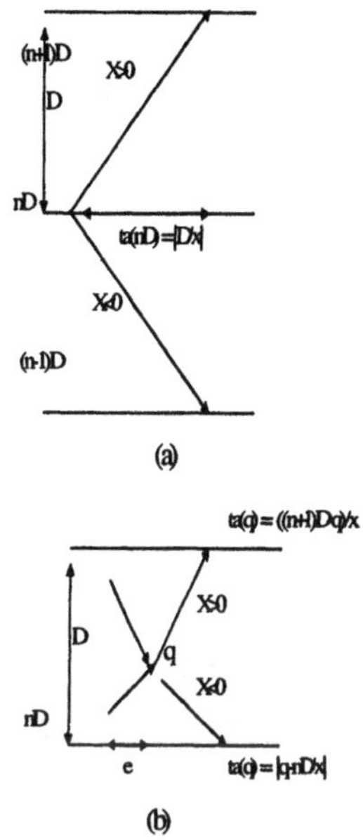

Figure 13. DEVS simulator of a Quantized Integrator

The Quantized DEVS integrator greatly reduces the number of transitions and output messages needed to simulate an integrator. It also reduces the message size from double to integer. Actually since the only possible transitions are to the upper or lower boundaries, only one bit to represent the binary valued set +1,-1 need be sent.

Recent developments have added to the stock of quantized integrators, see [15] for a review. Recent research in the representation of partial differential equations by DEVS is showing promising results also.

A.4 Multirate Sampled Systems

The CFSS formalism can be extended to directly represent Ordinary Differential Equations. An extension of the CFSS formalism to allow a direct representation of ODEs using the Euler method is described in [5]. Second order adaptive step-size Adam-Bashforth has been used in [8]. The representation of numerical methods on a sampled based formalism has the advantage of allowing its integration with *real* sampling-based systems like digital controllers [8]. We present a model of an integrator and a proportional controller that work with different sampling rates. The Euler Integrator is described by:

$$I = (X,Y,S,q_0,\rho,\delta,\Lambda)$$

where

$X = Y = \mathrm{R}$
$S = \mathrm{R} \times \mathrm{R} \times \mathrm{R}$
$q_0 = ((0,0,0.1),0)$
$\rho(x,y,ts) = ts$
$\delta((x,y,ts),x') = (x',y + x{*}ts,ts)$
$\Lambda((x,y,ts),e) = y + x{*}e$

The integrator has a timestep *ts* that is set to 0.1. The Euler method is used for the numerical integration and for the output extrapolation.

The proportional controller is described by:

$$P = (X,Y,S,q_0,\rho,\delta,\Lambda)$$

where

$X = Y = \mathrm{R}$
$S = \mathrm{R} \times \mathrm{R} \times \mathrm{R} \times \mathrm{R}$
$q_0 = ((0,4.0,2.0,0.2),0)$
$\rho(x,l,p,sr) = sr$
$\delta((x,l,p,sr),x') = (x',l,p,sr)$
$\Lambda((x,l,p,sr),e) = p{*}(x-l)$

The sampling interval of the controller is given by *sr* that is set to 0.2. Controller output is piecewise constant and is given by $p{*}(x\text{-}l)$, where *p* is the proportional coefficient (negative), *x* the sampled value and *l* a reference level.

The overall system is described in Figure 14. The proportional controller P samples from the integrator I. The Integrator samples from P at a different and independent rate.

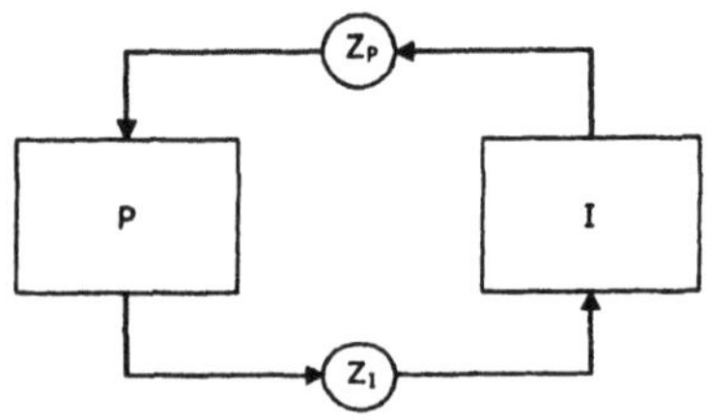

Figure 14. Integrator-controller network

References

[1] F.J. Barros. "Dynamic Structure Discrete Event System Specification: A New Formalism for Dynamic Structure Modeling and Simulation". *Proc. of the Winter Simulation Conference*, 781-785, 1995.

[2] F.J. Barros. "Modeling and Simulation of Dynamic Structure Discrete Event Systems: A Systems Theory Approach". PhD Dissertation, University of Coimbra, 1996.

[3] F.J. Barros. "A Framework for Representing Numerical Multirate Integration Methods". *Proc. of the 2000 AI, Simulation and Planning in High Autonomy Systems*, 149-154, 2000.

[4] F.J. Barros. "Modeling Mobile Software Agents in the CAOS Tool". *Proc. of DEVS Workshop, European Simulation Symposium*, 605-610, 2001.

[5] F.J. Barros. "Towards a Theory of Continuous Flow Models". *International Journal of General Systems,* 31(1): 29-39, 2002.

[6] F.J. Barros. "Modeling and Simulation of Dynamic Structure Heterogeneous Flow Systems". *Simulation,* 77(1): 552-561, 2002.

[7] F.J. Barros. "Towards Standard Interfaces in Dynamic Structure Discrete Event Models". *Proc. of AI, Simulation and Planning in High Autonomy Systems*, 170-174, 2002.

[8] F.J. Barros. "Harbor Modeling and Simulation: An Approach using Mobile Components". *Proc. of The International Workshop on Harbour, Maritime and Multimodal Logistics Modeling & Simulation*, 2002. (Accepted for publication.)

[9] M.S. Branick and S.E. Mattson. "Simulation of Hybrid Systems". In P.J. Antsaklis (ed.) *Hybrid Systems IV*, Lecture Notes in Computer Science, 1273: 31-56, 1997.

[10] A. Chow, B.P. Zeigler and D.H. Kim. "Abstract Simulators for the Parallel DEVS: Formalism". *Proc. of the Fifth Annual Conference on AI, Simulation, and Planning in High Autonomy Systems*, 157-163, 1994.

[11] Dept. of Defense. *High Level Architecture Interface Specification.* Version 1.0, US Defense Modeling and Simulation Office, 1996.

[12] N. Giambiasi, B. Escude and S. Ghosh. "GDEVS: A Generalized

Discrete Event Specification for Accurate Modeling of Dynamic Systems". *Transactions of the Society for Computer Simulation,* 17(3): 120-134, 2000.

[13] T.A. Henzinger. "The Theory of Hybrid Automata". *Proc. of the 11th Annual IEEE Symposium on Logic in Computer Science*, 278-292, 1996.

[14] Y.J. Kim and T.G. Kim. "A Heterogeneous Simulation Framework Based on the DEVS BUS and the High Level Architecture". *Proc. of the Winter Simulation Conference*, 421-428, 1998

[15] E. Kofman, J.S. Lee and B.P. Zeigler. "DEVS Representation of Differential Equation Systems: Review of Recent Advances". *Proc. of DEVS Workshop, European Simulation Conference*, 591-595, 2001.

[16] H. Praehofer. "System Theoretic Foundations for Combined Discrete-Continuous System Simulation". PhD Dissertation, Department of Systems Theory and Information Engineering, University of Linz, 1991.

[17] H. Sarjoughian, and B.P. Zeigler. "DEVS and HLA: Complimentary Paradigms for M&S?". *Transactions of the SCS,* 4: 187-197, 2001.

[18] D. C. Schmidt et al. "The Design of the TAO Real-Time Object Request Broker". *Computer Communications*, Elsevier Science, 21(4), April 1998.

[19] C. Thomas. "Interface-Oriented Classification of DEVS Models". *Proc. of 5th Annual Conference on AI, Simulation and Planning in High Autonomy Systems*, 208-213, 1994.

[20] B.P. Zeigler. *Theory of Modeling and Simulation.* Wiley, 1976.

[21] B.P. Zeigler and J.S. Lee. "Theory of Quantized Systems: Formal Basis for DEVS/HLA Distributed Simulation Environment". *Proc. of SPIE 11th Annual International Symposium on Aerospace/ Defense Sensing, Simulation and Controls: Enabling Technology for Simulation Science*, 3083: 49-58, 1998.

[22] B.P. Zeigler, H. Praehofer and T.G. Kim. *Theory of Modeling and Simulation.* Wiley, 2000.

11

A Simulation-Model Compiler for All Seasons[1]

Granino A. Korn
ECE Department, University of Arizona
7750 South Lakeshore Road, Chelan, #15, WA 68816
gatmkorn@aol.com

Summary. A new runtime simulation-model compiler quickly reads, compiles, and immediately solves vector differential equations and difference equations as well as scalar equations for dynamic-system models. Vector assignments compactly model nonlinear as well as linear matrix/vector relations for many problems in physics, control systems, neural networks, and fuzzy logic. The key feature is that the new vector differential-equation solver can replicate a dynamic-system model many times and then simulates hundreds of such models in a single simulation run, e.g. for Monte Carlo studies. Also, partial differential equations (e.g. for a heat exchanger) can be solved together with ordinary differential equations. Portable C code runs on UNIX or LINUX workstations. A fast machine-language version for personal computers (Windows 98 and NT) directly utilizes the Pentium floating-point hardware stack, which is automatically continued into memory.

1. Introduction

This report is organized as follows. Section 2 briefly reviews dynamic-system modeling with ordinary differential equations and the experiment-protocol program that exercises the differential-equation model to try the effects of different parameters and initial conditions. Section 3 continues the review to explain modern windows-type software designed for fast interactive experiments and exhibits a simple differential-equation-solving program.

Sections 3 and 4 then explore new territory. We introduce a novel vector compiler that reads vector and matrix equations, differential equations, and difference equations written much as in standard mathematics. A single

---

[1] Reprinted from *Simulation Practice and Theory (SIMPRA)*, v. 9, Granino A. Korn, A Simulation-Model Compiler for All Seasons, pp. 21-35, Copyright (2001), with permission from Elsevier Science.

simple program line can compile into many hundreds of equations or ordinary differential equations for statistical (Monte Carlo) studies, neural-network simulation, or partial differential equations. The compiled vector loops are unrolled, so that there is no runtime loop overhead. We exhibit several examples of useful simulations. Section 5 discusses compilation of submodels, again without any runtime overhead. Finally, Section 6 outlines the programming techniques needed to write such a fast (and portable) compiler for personal computers and UNIX workstations.

2. Dynamic-System Simulation

Simulation in teaching, research, and design is *experimentation with models.* To study dynamic systems in physics, chemistry, physiology, and aerospace, chemical, and control engineering, one usually models delayed interactions between physical *state variables* x_i with first-order ordinary differential equations (*state equations*)

$$(d/dt)x_i = F_i(t;\, x_1, x_2, \ldots;\, y_1, y_2, \ldots) \qquad (i = 1, 2, \ldots, n) \tag{1a}$$

where t is the system time, and

$$y_j = G_j(t;\, x_1, x_2, \ldots;\, y_1, y_2, \ldots) \qquad (j = 1, 2, \ldots, m) \tag{1b}$$

are "defined variables" typically representing intermediate results or system outputs. Some or all of the relations (1) may be formulated in vector/matrix form,

$$(d/dt)\, x = F(t;\, x, y) \qquad y = G(t;\, x, y) \tag{1c}$$

Other dynamic-system models (e.g. for digital control systems) employ difference equations, or both difference equations and differential equations.

A *simulation run* solves the system (1) to produce *time histories* $x_i(t)$ and $y_i(t)$ in successive time steps, starting at $t = t_0$ with given *initial values* $x_i(t_0)$. The run ends at t_0 + TMAX. The computer may also calculate functionals, typically performance measures such as a final value x_i(TMAX), or a mean-square error over the simulation run.

An *experiment-protocol* program sets parameters and initial conditions and then "exercises" the *simulation model* (1) in successive simulation runs. The experiment protocol can change parameters and/or initial conditions for successive simulation runs. Programmed or command-mode experiment protocols can also use simulation results to program in statistical or optimization studies.

3. Software for Interactive Modeling

Some popular simulation programs (SIMULINK™, VISSIM™) permit graphical entry of block-diagram simulation models and then *interpret* the program to permit interactive model changes. Such interpreted programs necessarily run relatively slowly, so that more efficient simulation software (AEGIS/ACSL™, DARE) translates screen-entered system equations (1) into Fortran, which is then *compiled and linked* together with appropriate library routines (e.g. for trigonometric functions). Unfortunately, compilation and linking imposes a 4- to 100-sec delay every time a program is entered or modified. These delays interrupt the experimenter's train of thought and are annoying, considering that you may want to modify your model hundreds of times in the course of a single modeling session.

DESIRE (Direct Executing Simulation in Real Time) simulation programs introduce a serendipitous compromise. We *interpret* only the experiment protocol, which programs only non-time-critical operations such as parameter setting between simulation runs. But differential-equation solving is time-critical in the extreme. Therefore, when the experiment-protocol program calls the first simulation run with the statement **drun**, an extra-fast run-time compiler *compiles* the right-hand sides of all model equations (1) (*DYNAMIC program segment*) directly into memory. A precompiled integration routine then executes immediately to produce solution graphs within less than 0.1 sec, an unnoticeable delay.

Properly designed software hides sophisticated program features from elementary users by defaulting to simple and frequently useful cases. So beginners quickly learn to solve simple differential equations with easily readable programs like

```
--DAMPED NONLINEAR HARMONIC OSCILLATOR
----------------------------------------------------------------------
TMAX = 2 | --                              simulation-run length
k = 30  | r = 1 | --                           model parameters
x = 1 | --                          initial value of x (xdot = 0)
drun | --                                 make a simulation run
----------------------------------------------------------------------
DYNAMIC
----------------------------------------------------------------------
d/dt x = xdot | --                                model equations
d/dt xdot = - k * x * x * x – r * xdot
dispt x  | --                               display x versus t
```

When you enter this program and then type **run** a solution graph appears at once. Note that the first 4 program lines specify the *interpreted experiment-protocol program,* which defines the experiment. The remaining lines specify the *compiled DYNAMIC program segment* that defines the simulation model. **dispt x** requests a runtime graph. The symbol | is a statement separator, and – is a comment delimiter. The DESIRE screen in Figure 1 shows a more interesting example.

Please note again that experimenters can immediately observe results of programmed or screen-edited model changes. The interpreted experiment-protocol operations are either *programmed* or *executed in command mode*, say for interactive debugging.[2] DESIRE/2000 solves up to 10,000 ordinary differential equations with fixed- and variable-step Runge-Kutta rules, or up to 200 differential equations with Gear and Adams variable-step/variable-order rules, using double-precision floating point numbers. More advanced users can access more and more features, such as user-defined functions of one or two variables and submodels, iterative multi-run studies, multiple DYNAMIC segments, and vector/matrix differential equations. The main subject of this report, though, is DESIRE's new vector compiler.

4. A New Vector Compiler

(a) Vector Assignments. Besides scalar linear or nonlinear equations such as

y = A * cos(w * t + phi) **d/dt xdot = - x * x * x - damping * xdot**

DESIRE/2000 can declare and manipulate one-dimensional arrays (vectors) and two-dimensional arrays (rectangular matrices). In particular, one-dimensional arrays of subscripted variables **x[k], y[k],** . . . and state variables **q[k]** represent vectors **x, y,** . . . **q**, say

$$\mathbf{x} \equiv (\mathbf{x[1], x[2], \ldots, x[n]})$$

Such arrays are dimensioned in the experiment-protocol program:

ARRAY x[n], y[n], z[n], r[m1], s[m2], . . . | STATE q[n]

Once declared, vectors can be used both in the interpreted experiment-protocol program and in compiled DYNAMIC segments. In particular, a DYNAMIC-segment model can specify ***vector assignments***,

Vector y = f(x, z, alpha, ...) - compiles into

y[i] = f(x[i], z[i], alpha, ...) (i = 1, 2, …, **n)**

where **x**, **y**, **z**, ... are all **n**-dimensional arrays (vectors), and **alpha** is a scalar. You can have any reasonable number of such **Vector** assignments with the same or different array dimensions **n**.

To generate the **n** vector components **y[i]**, a compiler loop, which causes no runtime overhead, reads the vector dimension **n** from the array data structure and compiles the assignment to the **i**th expression **f(q[i], x[i], y[I], alpha, ...)** **n** times, automatically incrementing the vector index **i** each time.

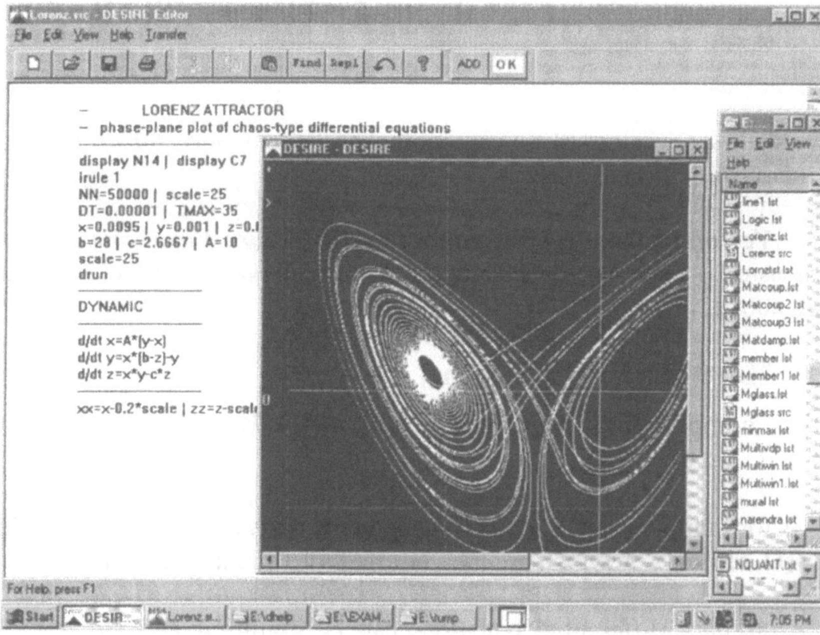

Figure 1. Solving the Lorenz-attractor Differential Equations.

This example generates a phase-plane plot for the classical Lorenz-attractor set of nonlinear differential equations. The user program was typed into an *Editor Window*. The interpreted experiment-protocol program sets parameters, initial conditions, and integration rule for the compiled differential-equation model in a DYNAMIC program segment.

Existing user programs can be loaded into an Editor Window from its File menu, but it is faster and much more convenient to drag them from the *Explorer Window* on the right. Several user programs, or different versions of the same program, can be open in *multiple Editor Windows*. This makes it easy to compare models. The selected program is activated by a mouse click on its **OK** button; then another click in the black *Run Window* then compiles and runs the simulation immediately.

The function **f(x[i], z[i], alpha, ...)** is the same for all **i**. It can freely use nonlinear expressions and unlimited nested parentheses. Scalars like **alpha** may be expressions containing literal numbers as well as symbolic variables. Here is a simple example:

Vector y = (1 – v) * (cos(alpha * z * t) + 3 * u)

- compiles into **y[i] = (1 - v[i]) * (cos(alpha * z[i] * t) + 3 * u[i])** (**i** = 1, 2, ..., **n**)

As we shall see, vectors like **z** can, moreover, be *matrix/vector products* **A * x** or *index-shifted vectors* **x{m}**.

(b) Vector Differential Equations and Difference Equations. Given an **n**-dimensional state vector **q** and **n**-dimensional vectors **x**, **y** declared as shown above,

Mat d/dt q = f(q, x, y, alpha, ...)

- compiles into **d/dt q[i] = f(q[i], x[i], y[I], alpha, ...)** (**i** = 1, 2, ..., **n**)

Our vector compiler again loops to generate **n** derivative components **f(q[i], x[i], y[I], alpha, ...)**, which are then inserted into an integration routine.

The **n** ordinary differential equations can be nonlinear and are integrated automatically, with a choice of integration rules. The experiment protocol can assign initial values for the state variables **q[i]** or accept the default value 0.

A third vector operation,

Delta x = f(x, y, alpha, ...)

- compiles into **x[i] = x[I] + f(x[i], y[i], alpha, ...)** (**i** = 1, 2, ..., **n**)

compactly implements a vector difference equation useful e.g. for recursive summation.

(c) High Computing Speed for Parameter Studies and Monte-Carlo Simulation. Note, at this point, that these vector operations cause ***no runtime loop overhead.*** They simply compile into multiple scalar assignments (loop-unrolling compiler). DESIRE/2000 currently handles up to 10,000 first-order differential equations.

With a suitable parameter array filled by the experiment-protocol program, we can exercise replicated a models with different parameter values in a single simulation run and tabulate or cross-plot results. But a more interesting application of the **Vector** and **Mat d/dt** operations is ***Monte Carlo simulation by model replication.***[1] As a simple example, we can model **n** = 1000 cannonball trajectories with air resistance [2] and random wind forces:

Vector v = sqrt(xdot * xdot + ydot * ydot) | -- computes **n** airspeeds **v[i]**
--
Mat d/dt x = xdot | **Mat d/dt y = ydot** | -- 4**n** state equations for **n** models
Mat d/dt xdot = - R * v * xdot + ran()
Mat d/dt ydot = - R * v * ydot - g

A single simulation run now generates all 1000 random-input trajectories and readily computes runtime statistics.[1] Since there is no loop overhead, a Pentium-class personal computer produces the equivalent of more than 1000 repeated trajectories in one simulation run (Figure 2). With an 800-MHz Pentium III this takes about 0.3 sec, compared to over 1 sec for 1000 repeated Monte Carlo runs.[1] This permits very fast interactive studies of the effects of parameter changes *on population statistics,* such as mean-square guidance-system errors.

5. Vector/Matrix Products And Index-Shifted Vectors Implement Interconnections Between Model Variables

(a) Going beyond simple Model Replication. With simple model replication, each vector equation or differential equation, say

Mat d/dt q = f(q, x, y, alpha, . . .)

relates only vector components elements **q[i]**, **x[i]**, **y[i]**, . . . all labeled with the same index **i**. But more sophisticated and useful models must also be able to relate different components of the same vector, and also components of different vectors, which need not have equal dimensions.

(b) Matrix/vector Products Relate Different Vectors. Any vector, say **z**, on the right-hand side of a **Vector, Mat d/dt, or Delta** operation can be a ***matrix-vector product*** **A** * **v**, where **v** can be defined by another **Vector** assignment.[†] Here **A** is a rectangular matrix declared with

ARRAY A[n, m]

and "filled" with array elements **A[i, k]** assigned by the program.
Simple vector assignments like **Vector y = A * x conveniently** implement geometrical coordinate transformations, such as rotations. A more interesting example is the vector/matrix representation of a simple nonlinear ***two-layer neural network***:

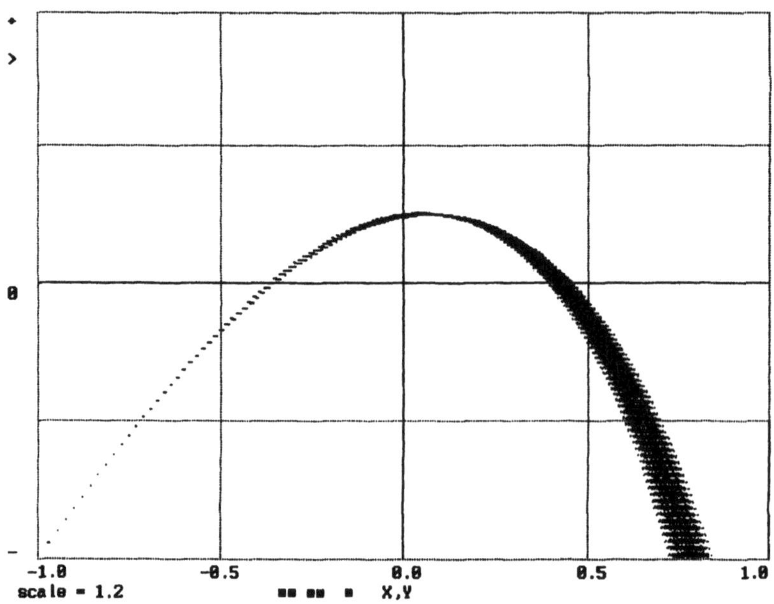

Figure 2a. 500 cannonball trajectories with noisy wind forces were run in succession for this repeated-run Monte Carlo simulation. On a 400-MHz Pentium II, this took 1.95 sec with the display turned off.

[†] If **v** is identical with assignment target of a **Vector** or **Delta** operation, the program reminds you that an intermediate vector assignment **Vector z = A * v** is needed, just as it would be in Fortran or C. There is no such problem in **Mat d/dt** operations, since the differential-equation solver automatically assigns a hidden intermediate variable for each derivative.

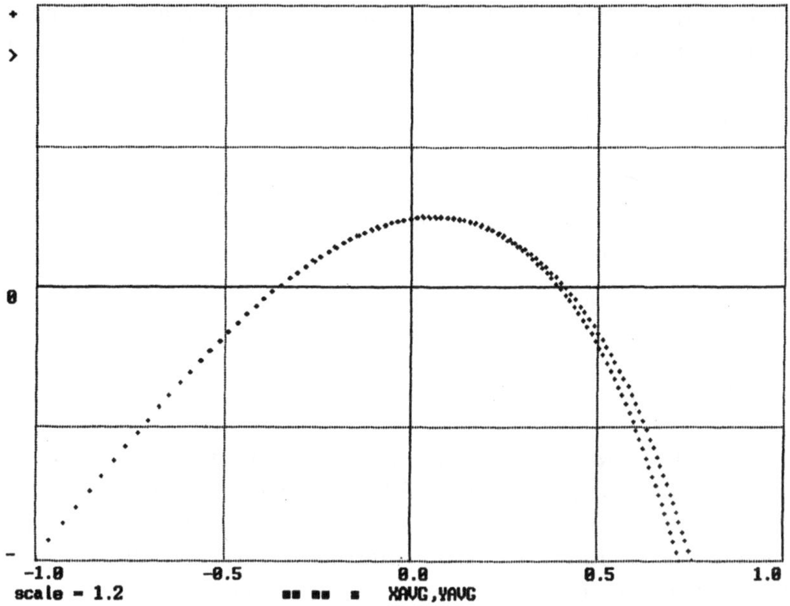

Figure 2b. Replication of 500 trajectory runs lets you compute Monte Carlo population averages in a single simulation run (see text), so that the effects of model-parameter changes *on statisitics* can be studied. The figure shows average trajectories for two samples of n = 500 replicas of the cannonball model. Each simulation run (the equivalent of 500 repeated runs) took 0.25 sec on he 400-MHz Pentium II. Note that the sample averages xavg and yavg are themselves random variables.

ARRAY layer1[n], layer2[m], layer3[r], bias1[n], bias2[m], W2[m, n], W3[r, m]

....

Vector layer2 = tanh(W2 * layer1 + bias1)
Vector layer3 = tanh(W3 * layer2 + bias2)

<u>This automatically compiles into multiple equations,</u>

$$\text{layer2[i]} = \tanh\left(\sum_{j=1}^{n} \text{W2[i, j] layer1[j]} + \text{bias1[i]} \right. \qquad (i = 1, 2, \ldots, m)$$

$$\text{layer3[k]} = \tanh\left(\sum_{j=1}^{m} \text{W3[k, j] layer2[k]} + \text{bias2[k]} \right. \qquad (k = 1, 2, \ldots, r)$$

Vector and **Mat d/dt** operations can model classical perceptrons and backpropagation networks, but also sophisticated pulse-coupled (oscillating) neurons,[4] as well as recurrent neural networks incorporating feedback and time delays.[3] For this purpose, DESIRE can also compile matrix equations and difference equations used, for instance, to update neural-network connection-weight matrices.[3] Special operations implement competitive neuron layers.[3] Figure 3 shows two examples.

The same easily readable vector/matrix notation also produces ***compact models of linear or nonlinear control systems*** with multiple inputs and/or outputs. A good example is the familiar classical state-transition model

Mat d/dt q = A * u + B * u
Vector r = C * q + D * u

which relates outputs **r[i]** and inputs **u[k]** of a linear control system.

(c) Index-shift Operations. Other important applications relate only a few different components of the same or different vectors, say **x[i]**, **x[i + 1]**, **x[i – 2]**, **y[i-2]**, ... We permit any vector on the right-hand side of a **Vector**, **Mat d/dt**, and **Delta** operation to be an ***index-shifted vector*** denoted by **u{m}**, with components **u[1 + m]**, **u[2 + m]**, ... Specifically,

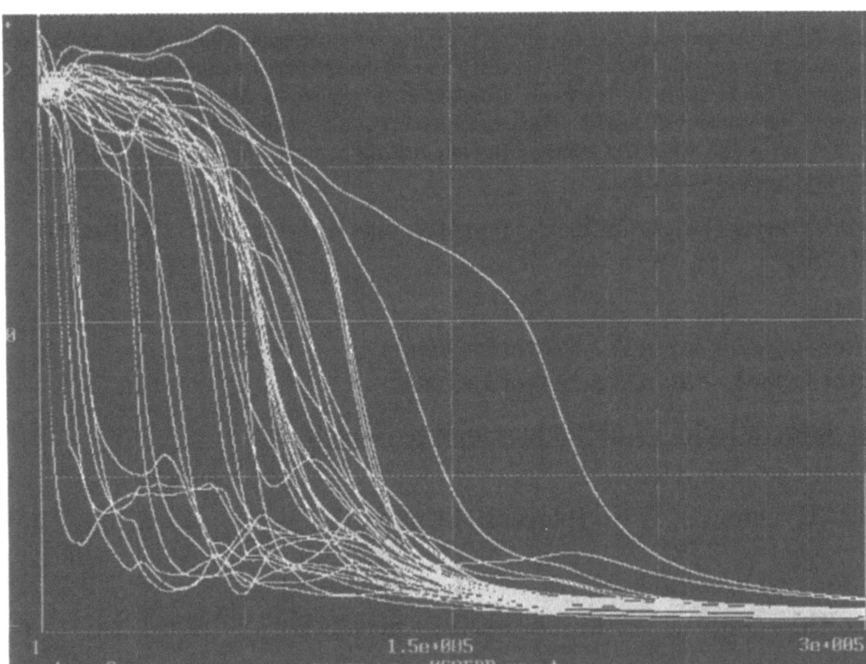

Figure 3a. A backpropagation network learning 32 patterns simultaneously. All 32 squared pattern errors are plotted against time.

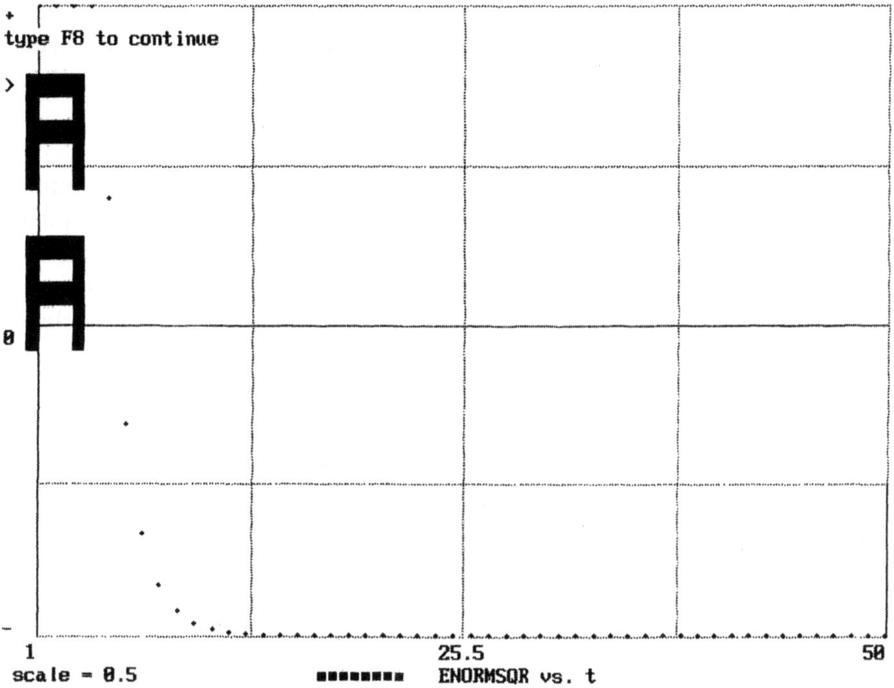

Figure 3b. A character-recognition network learning the pattern for the letter A (based on Ref. 3). The square of the pattern error is plotted against time to show how the error converges.

Vector y = sin(x{m}) compiles into[‡] **y[i] = sin(x[i + m])** (**i** = 1, 2, . . . , **n**)

The index shift **m** is an integer expression set at compile time. When **i + m** < 1 is > **n** then **x[i + m]** is simply replaced by 0.

As an example, we can declare an **n**-dimensional vector **x** with **ARRAY x[n]** and program

output = x[n] | Vector x = x{- 1} | x[1] = input

[‡] If **v** in **v{m}** is identical with assignment target of a **Vector** or **Delta** operation and **m** > 0, the program reminds you that an intermediate vector assignment **Vector z = v{m}** is needed, just as it would be in Fortran or C. Again, there is no such problem in **Mat d/dt** operations, since the differential-equation solver automatically assigns a hidden intermediate variable for each derivative.

to shift a given scalar input **input(t)** through **x[n]**, **x[n –1]**, . . . , **x[1]** and then into the scalar output **output(t)** in successive simulation time steps. This models a delay line useful, for instance, to shift a time function into transverse filter or predictor network (serial-to-parallel conversion).

DESIRE's index-shifting feature was originally designed to model recurrent neural networks incorporating time delays, lateral inhibition, or adaptive predictors.[3] More recently, index-shifting has also neatly implemented space-coordinate differentiation **for *interactive Method-of-lines solution of partial differential equations*** (Figure 4).[5,6] This lets you directly include models of heat exchangers, simple chemical reactors, etc. in process-control simulations modeled with ordinary differential equations.

6. Desire Submodels Compile Without Runtime Overhead

Modeling languages are further extended with *user-defined functions* and *submodels*. Submodels implement special function blocks or complete sets of equations and differential equations

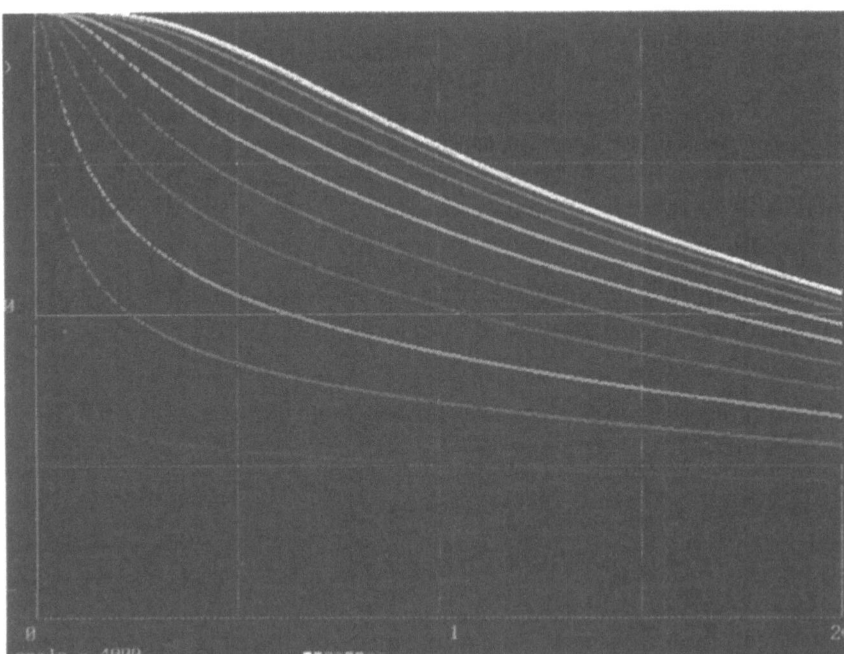

Figure 4: Solving a Partial Differential Equation by the Method of Lines.[6]

For a Method-of-lines solution of the partial differential equation (heat conduction in a rod, temperature *u* is a function of *x* and *t*)

$$\partial u/\partial t = uxx \qquad (ux \equiv \partial u/\partial x,\ uxx \equiv \partial ux/\partial x)$$

we declare **N**-dimensional arrays of corresponding node values of *x*, *u*, *ux*, and *uxx* :

ARRAY x[N], u[N], ux[N], uxx[N]

A numerical-differentiation submodel uses vector-shift operations to produce the array ux from *u* and uxx from *ux*, with modified difference formulas at the boundaries:[5]

```
        MACRO  DDx(n, b, v, vx)
          --
Vector vx = b * (v{1} - v{-1}) | --              numerical differentiation
    --
vx[1] = b * (- 3 * v[1] + 4 * v[2] - v[3]) | --   differentiation at boundaries
vx[n] = b * (3 * v[n] – 4 * v[n - 1] + v[n - 2])
end
```

The experiment protocol program sets parameters and initial conditions and then calls a DYNAMIC program segment to solve ordinary differential equations. The boundary condition for x[N] is a nonlinear heat-radiation condition.

```
DYNAMIC
-------------------------------------------------
invoke DDx(N, B, u, ux) | --                        differentiate u to get ux
ux[1] = 0 | ux[N] = E * (UA4 -  u[N]^4) | -- set boundary values for ux
invoke DDx(N, B, ux, uxx) | --                      differentiate ux to get uxx
Mat d/dt u = uxx | --                               N ordinary differential equations
```

The program displays solution values **u[i]** for different **x[i]** as functions of the time *t*. Representing, say, a small servomechanism or a neural network for convenient reuse. Submodels are much more than a convenient shorthand notation: they are abstractions that make it easier to understand models. A DESIRE ***submodel declaration***

```
MACRO  F(x$, y$, . . . )
  prototype code lines
  . . . . . . . . . . . . .
  end
```

includes DYNAMIC-segment lines referencing dummy variables **x$, y$**, . . . , which can be scalars, vectors, or rectangular matrices. When a DYNAMIC program segments *invokes* the submodel with

INVOKE F(x, y, . . .)

the invocation reproduces the submodel code with new variables **x**, **y**, . . . corresponding to the dummy variables **x$**, **y$**, . . . (Figure 4). Interestingly, DESIRE submodels are *not* implemented as text-reproducing macros, and they do *not* invoke runtime subroutines. They generate *compiler subroutines* that produce runtime code *without any overhead for function calls.*

7. How It Is Done: Compiler Implementation For Personal Computers And Unix

(a) In-line and Threaded-code Implementations. The earliest DESIRE runtime compilers produced true in-line machine code for the DEC PDP-11and VAX, and then for Intel personal computers. The compiler front end was written in Pascal, but the code generator employed machine-specific assembly language to produce the highest possible execution speed. For the same reason, we used single-precision code to compute derivatives, with double-precision integral accumulation. A machine-language runtime library allowed simulations to outrun personal-computer Fortran.

This was very efficient, but would be prohibitively expensive on RISC-type UNIX workstations, especially since each CPU type would require a new machine-language code generator. For this reason, more recent DESIRE compilers are written in ANSI-C and generate threaded code (virtual-machine code) rather than in-line code. This also makes large vector programs much more manageable. The new compilers are *portable*: they can be recompiled, with minimal changes, for different computers and operating systems. The system is currently supported with the Windows interface shown in Figure1 (Windows 98, Me, NT, and 2000), and as a console program displaying color graphics under SOLARIS™, LINUX, ans MS-DOS. Experimental versions have also run under AIX™, VMS/DECwindows™, and OS/2. We employ double-precision (64-bit) floating-point code throughout, not just for integral accumulation.

Our compilers produce efficient modified stack-machine code, i.e. they generate "add memory to stack top", "multiply memory into stack top" as well as simple push and pop operations. Thus the expression assignment

x = a * b + c * d

produces optimal "stack-machine code",

```
push   a
multiply stack top by b
push   c
multiply stack top by d
add top stack registers
pop    x
```

(b) ***Portable C Implementation.*** The portable threaded-code compiler compiles an array

void (*COMPILE[N])()

of successive function pointers pointing to pseudoinstructions implemented as C functions **COMPILE[ci]()**, and also an array of successive integer data pointers **dptr[M]** for these functions. In effect, the corresponding array pointers **ci** and **di** function as the program counter and address register of a *virtual machine* whose instructions are function calls.

In this way, **push a** and **pop a** compile into function pointers which point to

```
data[- -ss] = data[dptr[di++]] ;
data[dptr[di++]] = data[ss++] ;
```

Similarly, **multiply stack top by b** points to

```
data[ss] * = data[dptr[di++]] ;
```

The array **data** contains double-precision floating point data. The C variable **ss** serves as the stack pointer of the virtual machine. Functions like **sin(x)** are returned on the virtual-machine stack.

(c) ***Machine-language Hardware-stack-machine Code Generator for Personal Computers.*** Our personal computer (Windows) program is largely identical with the UNIX program, except that now the virtual machine is implemented in Pentium-optimized *machine code. The Pentium floating-point processor serves directly as a true stack machine, and the virtual program counter and data address register are real hardware registers.* What is more, assembly language lets us to replace function calls and returns with much more efficient jumps via the Pentium register **ECI**, which conveniently serves as a hardware program counter and is incremented after each jump. We show an example in the Appendix.

The Intel processors used in personal computers have a double-precision floating-point hardware stack with appropriate instructions. But this stack has only 8 directly accessible registers, of which two must be reserved to generate functions such as exponentials. At this point, DESIRE uses a special trick. The compiler maintains a stack counter and compiles the normal processor-stack instructions listed above as long as the stack depth does not exceed 6 registers. But when a long expression requires more than 6 stack registers, we automatically compile modified **push** and **pop** routines that continue the processor stack into memory with machine-language **push register** and **pop into register** instructions. Since stack counting is done by the compiler it imposes no runtime overhead.

(d) Execution Speed. Vector-compiled Monte Carlo simulation was 2 to 3 times faster than the repeated simulation runs needed without vector compilation.[2] This is true because of the saved vector-loop overhead. Since the loop-unrolled code is necessarily larger, the exact speed advantage depends on the amount of L2 cache available to the program.

Threaded-code execution incurs the overhead of indirect data references and repeated function calls. Our special machine language program reduces the function-call overhead and generates true hardware stack-machine code. In general, threaded machine-language code was found to be only 2 to 3 times slower than true in-line machine code. Using portable C code instead of machine language slows you down by another factor of 2 to 3.

With a 400-MHz Pentium II, a simple three-dimensional flight simulation required

— 0.07 sec with true in-line compilation of single-precision machine code
— 0.26 sec with threaded-code compilation of double-precision machine code
— 0.74 sec with threaded-code compilation of double-precision C code.

Newer personal computers (Pentium III or Pentium 4 processors with Rambus memory, or ATHLON processors) and high-end UNIX workstations are 3 to 4 times faster. It seems fair to use some of the processor speed to pay for code portability and maintainability.

References

[1] G.A. Korn. *Fast Interactive Monte Carlo Simulation of Dynamic Systems.* Model Replication (*submitted*), 1999.

[2] G.A. Korn. *Dynamic-system Simulation with Microsoft Windows 95 and NT.* Gordon and Breach, London, 1998.

[3] G.A. Korn. *Neural Networks and Fuzzy-logic Control on Personal Computers and Workstations,* MIT Press, Cambridge, MA, 1995.

[4] J.L. Johnson and M.L. Padgett. "Pulse-coded Neural Network (PCNN) Models and Applications". *Trans. IEEE on Neural Networks*, 10: 480-498, 1999.

[5] W.E., Schiesser. *The Numerical Method of Lines.* Academic Press, New York, 1991.

[6] G.A. Korn. "Interactive Solution of Partial Differential Equations by the Method of Lines". *Mathematics and Computers in Simulation,* 1644, 1-10, 1999.

[7] *Pentium III Programmer's Manual*, Intel Corporation, Santa Clara, CA, 1998.

References 2 and 3 contain educational versions of the simulation software described in this report, but a more up-to-date version can be downloaded without cost from *http://members.aol.com/gatmkorn.*

APPENDIX: Fast Threaded-code Implementation on Personal Computers

In Pentium (or Athlon) assembly language, push **a** now produces

```
MOV  EAX, DWORD PTR _dptr[EDI * 4]    ; load 4-byte data address
INC  ESI                              ; increment program
counter
FLD  QWORD PTR  _data[EAX * 8]        ; push a  (8 bytes)
INC  EDI                              ; increment data pointer
JMP  DWORD PTR _COMPILE(ESI * 4)      ; next pseudoinstruction
```

and **multiply stack top by b** points to

```
MOV  EAX, DWORD PTR _dptr[EDI * 4]    ; load 4-byte data address
INC  ESI                              ; increment program
counter
```

```
FMUL   QWORD PTR  _data[EAX * 8]       ; multiply stack top by a  (8
bytes)
INC    EDI                             ; increment data pointer
JMP    DWORD PTR _COMPILE(ESI * 4)     ; next pseudoinstruction
```

The hardware register **EDI** is the virtual address register. The vital **push** and **pop** operations execute on the actual hardware coprocessor stack. Note that we used scaled indexing. **INC EDI** and **INC ESI** were transposed to optimize Pentium pipeline operation.[7]

12

Simulation Languages and Applications

Ralph C. Huntsinger
Emeritus Professor of Computer Science
Emeritus Professor of Mechanical Engineering
California State University, Chico, CA 95929
drralph@ecst.csuchico.edu

Summary. This chapter describes past and present continuous and discrete systems digital simulation languages and some applications. The emphasis is on the developing history and the evolution of simulation languages.

1. Introduction

Professor Walter Karplus was one of the pioneers in the use of simulation languages to solve real world engineering problems. In this chapter a review of the past history of simulation languages will be covered. First, continuous systems simulation languages for the solution of non-linear, non-homogeneous, simultaneous, high order, ordinary differential equations will be covered. Second, discrete systems simulation languages for the solution and analysis of waiting line systems will be covered. At the end of this chapter selected applications will be listed.

2. Continuous Systems Simulation Languages

One of the very early continuous systems simulation languages (CSSL) was MIDAS which stood for modified integration digital analog simulator. It was such a great language that the creators named it after King Midas the king with the golden touch. MIDAS was block oriented like an analog computer diagram and then an improved version was developed called MIMIC that was equation oriented. These were both developed at the Wright Patterson AFB. Bob Brennan of IBM wrote a block diagram oriented language called Pactolus which was changed about 9 months later by IBM marketing to 1130 CSMP (Continuous Systems Simulation Language). You see King Midas had some problems with the golden touch. When he shook hands with his friends they did not talk back any more and his bed was very hard, etc. Well, to get rid of the golden touch the king was told to go into the river Pactolas, so Pactolas was the language that made MIDAS lose its "golden touch". Block diagram languages were very popular with the old analog computer crowd as they looked just like analog computer diagrams and in fact were first used as digital-analog check scaling programs for analog computers. The first digital computers were quite slow and the analog computer was still king for a while. The simulation language engines are numerical method techniques for solving ordinary differential equations. Some languages have a better selection of numerical methods than others and some handle discontinuities much better than others do. This will be commented on later in the chapter. The names of CSSL's are interesting in themselves so here are some fun examples. At Bradford University in England a group developed a simulation language for the chemical engineering students and it was called BEDSOCS (Bradford Educational Simulation of Continuous Systems). Of course, it is very cold and damp most of the time in that part of England so folks wear socks to bed to keep their feet warm, hence BEDSOCS. The language DESIRE (Direct Execution Simulation in Real time) was written by Professor Korn from the University of Arizona. He published an early version of the language in SIMULATION Journal called "Early Desire". Sociologists and Psychologists were busy looking up his article only to find differential equations and computer programs. This was a bit of a shock for most of them. Professor Korn also wrote a simulation language called DARE (digital analog replacement) which was also written in BASIC and called B-Dare. The U.S. Navy China Lake research facility modified DARE to fit their own requirements and called it CLOD (China Lakes Own DARE). IBM was in the beginning of the development of simulation languages with their 360/CSMP (continuous systems modeling program) which was an equation oriented language that was used by many people in the early days of simulation languages. The Siemens Corporation ported CSMP to their Siemens main frame computers and called it SICOS (Siemens Continuous Simulator). The Universities in Belgium did not use the german translated manuals and instead used the english CSMP manuals to run the program on the Siemens Computers in Belgium. Simulation languages were everywhere and a great example was EZQ that ran on an Apple IIe . After the success of CSMP IBM continued the development with DSL/DP (Digital Simulation Language/ Double Precision) by Mr. Wy Mun Syn who was working in the simulation group at IBM lead by Dr. Horace Mendez. IBM marketing then changed the name to DSL/VS (Dynamic Simulation Language/ VS Fortran) to fit the new VS version of Fortran that was the base language for DSL/VS. The NASA Marshall Space Flight Center then modified the original DSL/DP to produce MARVES (the Marshall Vehicle Engineering Simulator) which had the advantage of having Dr. Edwin Fehlberg's integration routines as the numerical analysis engine. The Fehlberg routines are in almost all simulation languages now but MARVES was the first language to have them. Electronic Associates Inc. also came up with their own language called HSL (Hytran

Simulation Language) which was used primarily as a check scaling language for their line of Hybrid Computers. This was later replaced by a regular equation oriented simulation language called AD-SIM (Applied Dynamics Simulation). The Boeing Corporation got into simulation languages in a big way with their EASY-5 (Engineering Analysis of Systems version five). This was a modular block oriented language with sub-blocks and graphic animation. This was a language for large problems and needed a workstation level computer to run on. The language was marketed by Boeing Computer Services for several years. Another block oriented language was XANALOG that combined the software with fast hardware boards to give fast execution times. A language that was developed by Roger G.E. Franks at Du Pont to go with his text book on "Mathematical Modeling in Chemical Engineering" was distributed as a deck of punched cards (those were the days...) that was all written in Fortran and was a very easy language for students to use. In the old days many students cut their simulation teeth on the language that we all called R-Gef. Some languages appeared and disappeared along with the computer companies that they were attached to. One good example is SL-1 (Simulation Language one) that ran on the old Sigma-3 and Sigma-5 computer systems. IBM was put into the simulation language business again when Dr. Manesh Shah and a Co-Op student wrote a language called PC-ESP (Personal Computer Engineering Simulation Program) and then Dr. Shaw wrote a book published by Prentice Hall called "Engineering Simulation: Tools and Applications using the IBM PC Family". The language was first written in BASIC and then a larger version was written in Fortran. A very fine block oriented simulation language was modified and marketed in the USA by Walter Reynolds and he named it TUTSIM as it was first written at the University of Twente in the Netherlands. An entire book was written based on this language by Nicholas Karayanakis titled "Computer-Assisted Simulation of Dynamic Systems with Block Diagram Languages". After TUTSIM was no longer supported in the USA, the folks at the University of Twente control lab rewrote and expanded the language and named it 20-Sim (Twente Simulation). This is a very powerful PC-Windows based software package that also supports Bond-Graph modeling of electro-mechanical systems. Another high level package is CAMP (Computer Aided Modeling Program) written by Professor Jose Grande of Sacramento State University. The graphical interface allows the engineer to construct Bond-Graph models on the screen and then converts the diagrams into ACSL or CSSL-IV simulation programs. ACSL (Advanced Continuous Systems Language) and CSSL-IV (Continuous Systems Simulation Language version four) are the latest two versions of a long family of related languages. The past languages in the family are CDC's CSSL-III, NavSim, and RSSL (Raytheon System Simulation Language). All of these languages have many of the same commands and macros and at one time shared most of the same root Fortran code. These are preprocessor languages and produce either Fortran or C code that runs for the execution phase of the simulation. This makes them very fast. Another language that is a modern language and also can produce Fortran execution code is ESL (the European Space Agency Simulation Language). ESL is both an equation oriented language and a block diagram oriented language. The equation oriented form uses an "Ada" language format and the block diagram form will write code in the equation oriented from which can then be run on an interpreter for debugging or can be used to produce Fortran for fast execution. The best feature of ESL is its ability to handle discontinuities very well. ESL uses a procedure developed by Professor Roy Crosbie who is now at California State University, Chico. A language that uses signal flow graphs or Bond-Graphs for simulation is PASION developed at the Universidad Panamericana in Mexico City by Professor Stan Raczynski. Several languages have been written with the Personal Computer in mind. An example is ISIM (Interactive Simulation) first developed for the CP/M operating system. An article in BYTE magazine in May of 1984 by Professor Crosbie shows how to use ISIM to solve the classic "Rocket" problem. Other examples are PSI from the Delft University of Technology, ISL, and ISIS. Some simulation languages were written with special users in mind. SAAM-II (Simulation Analysis and Modeling version two) was created by Dave Foster's team at the University of Washington for BioMedical Engineering applications and STELLA was written for the Biology community. Last on the scene are Powersim from the Norwegians and Simulink on Matlab. The author of this chapter collects simulation languages for his simulation courses. He has written and executed programs in the following Languages.

1) CSSL-IV (Continuous Systems Simulation Language 4)
2) DSL/VS (Dynamic Simulation Language, IBM)
3) ACSL (Advanced Continuous Systems Simulation Language)
4) TUTSIM (Block Oriented Language for PC's)
5) EASY-5W (Engineering Analysis System 5 by Boeing Co.)
6) PC-ESP (PC-Engineering Simulation Program, IBM)
7) DESIRE (Direct Executing Simulation Language in Real Time)
8) ISIM (Interactive Simulation Language for the PC)
9) DARE-P (Differential Analysis Replacement - Portable)

10) CSSL-III (CDC Original Version of CSSL-IV)
11) ESL (European space agency Simulation Language)
12) ADSIM (Applied Dynamics Simulation Language)
13) CSMP-III (Continuous Systems Modeling Program, IBM)
14) 1130 CSMP (Block Oriented Language by IBM)
15) SYSL/M (System Simulation Lang: CSMP-III on the PC)
16) CAMP (Computer Aided Modeling Program)
17) MARVES (Marshal Vehicle Engineering Simulation System, NASA)
18) HSL (Hytran Simulation Language, Electronics Assoc., Inc.)
19) MIDAS (Modified Integration Digital Analog Simulator)
20) MIMIC (Simulation Lang. Developed at Wright Pat. A.F.B.)
21) RGEF (Du Pont Engin. Simulation Prog.)
22) SL-I (Simulation Language One)
23) PSI (Block Oriented Control Systems Lang. from Delft Univ. of Technology)
24) SICOS (Siemens Company Simulation of Continuous Systems)
25) XANALOG (ICON Oriented Simulation Lang. with Iconic Graphics)
26) BEDSOCS (Bradford Educational Simulation of Continuous Systems)
27) DYNAMO (System Dynamics Modeling Program)
28) NAVSIM (U.S. Navy Version of CSSL-III)
29) ENPORT (Linear Systems Bond Graph Analysis Program)
30) EZQ (PC and Apple II Block Diagram Simulation Program)
31) COSMOS (Continuous and Discrete Systems Modeling Program from the Netherlands
31) PASION-CMG (Continuous Model Generator by S. Raczynski)
32) 20-SIM (Block Diagram Simulator from the Univ. of Twente Control lab)
33) SIMULINK (from The MATH WORKS Inc. - MATLAB)

Why are there so many continuous systems simulation languages? I think that one reason is the NIH (Not Invented Here) factor. Of course everyone thinks that they can build a "better mousetrap" and if the market niche is there then "money motivates".

3. Discrete Systems Simulation Languages

Discrete systems simulation languages solve queuing problems (waiting line systems) and are used for the analysis of manufacturing systems, computer operating systems, etc. These languages are productivity increasing software packages. For example a simulation of a computer memory management system written in C++ with 1800 lines of code was rewritten in a discrete systems simulation language with 115 lines of code and obtained the same results. In the beginning there was Geoffery Gordon of IBM and his development of GPSS/360 in the early 70's as a block diagram simulation language for discrete stochastic systems. This language was very easy to learn and enjoyed great success with the user community. It evolved into the GPSS-V (General Purpose System Simulator version five) that became the standard for many other simulation languages. GPSS-V was an interpreted language and did everything with integers. Jim Henrikson of Wolverine Software wrote a version of GPSS-V with extensions that compiled, used floating point arithmetic and was very fast. He called his version GPSS/H and this supported his new company. The first versions ran on main frame computers and the new versions run on PC's. Wolverine Software has evolved GPSS/H into a newer product called SLx (Simulation Language Extended) that goes beyond the limitations of GPSS/H. Other versions of GPSS have been developed and an interesting one was GPSS-Fortran developed in Germany by Professor Bernd Schmidt. This was a set of Fortran subroutines that matched the functionality of each of the standard GPSS blocks. Other block oriented languages are SLAM-II (Simulation Language for Alternative Modeling version two), SIMAN, and GPSS-PC. Siman has evolved into "Arena" and SLAM-II has evolved into "SlamSystem". The non-block oriented languages started with Simula-67 with class Simulation. Simula was the first object oriented programming language and its primary first use was discrete system simulation. Professor Graham Birtwhistle wrote a front end program in Simula of about 2400 lines of code and called it DEMOS (Discrete event Modeling on Simula). DEMOS made Simula look like GPSS and made it very easy to write simulation programs using Simula. Another company that was active in discrete system simulation is CACI Products. They market Simscript-II.5 and MODSIM-III which are both high level languages. Simscript sort of looks like a cleaned up Fortran with special features and is fairly easy to use while MODSIM-III is a more modern and more powerful language. Promodel is another fine language for manufacturing applications. Many of these languages have animation packages integrated into the software package. Animation in a discrete systems simulation is not that useful for analysis as these languages all describe the performance of a system at steady state and the summary statistics give all of the valuable information. Animation is fun to watch and is good for showing the Generals and Vice Presidents that you are doing work in the simulation area. GPSS/H has animation with its PROOF package, Simon has animation with Arena, etc. The real use of animation is to help the simulationists keep their jobs. The PASION simulation language from the Universidad Panamericana has a discrete systems simulation module that has an excellent animation package that is very easy to use. Last but not least there is a company named Alphatech that markets a simulation package called ALPHA/Sim. This language is a bit different as it uses Petri nets as the elements of the language. There is a large literature on Petri nets and this language enables those users that like this procedure to use a high level simulation language. There are two ways of doing discrete simulation, the event method and the process method. In the event method one tells the language when to start a task and later on tells the program to stop the task. If you forget to stop a task you get very weird results. In the process method one tells the language when the task will start and also at the same time tells the language how long the task will last. Almost all modern discrete system simulation languages use the process method for simulation. Many users of discrete systems simulation languages are Industrial Engineers interested in manufacturing systems and Computer Network managers interested in communication systems. Performance evaluation of computer systems is also an area of interest. The following are the languages that the author has used to write simulation programs.

1) GPSS-V (General Purpose System Simulator, IBM)
2) GPSS/H (Wolverine Software Version of GPSS-V)
3) SIMULA (Norwegian Computer Center Language)
4) DEMOS (Discrete Event Modeling on Simula)
5) SLAM-II (Simulation Language for Alternative Modeling)
6) GPSS-FORTRAN (Univ. of Erlangen-Nuernberg Development)
7) SIMSCRIPT-II.5 (CACI High Level Simulation Language)
8) PASSIM (Pascal Simulation Language from Wala Wala Univ.)

9) SIMPLE_1 (PC Based Discrete Systems Simulation Language)

10) Q-GERT (Queuing Graphical Evaluation Language)

11) Sim++ (Time Warp language by Jade Simulations International)

12) GPSS-PC (GPSS version by Minuteman Software Inc.)

13) PASION (Delphi based Process and Event-Oriented language)

4. Applications

John McLeod, the founder of the Society for Modeling and Simulation, is quoted as saying that "simulation is a hopeful fake when the real thing is just too much". Dr. Ragnar Nilsen, the owner of Simulation Services that markets CSSL-IV, printed a bumper sticker that said "Reality is a Special Case, Simulation goes beyond!". The author has been quoted as saying "Simulation is fun". Continuous systems simulation languages have a wide applicability in all fields involving dynamically varying phenomena including but not limited to the following areas. The areas are Aerospace Vehicle Dynamics, Automatic Control Systems, Physiological Systems, Nuclear Engineering, Physics, Chemical Processes, Automotive Design, Thermodynamics, Electrical Engineering, Environmental Systems, Process Control, Maritime Systems, Training Systems, Fluid Dynamics, Education, Biomedical Systems, Power Utilities, Mechanics, Applied Mathematics, Robotics, Agricultural Systems, and the list goes on and on. Discrete systems simulation solves problems in Manufacturing Systems, Business Systems, Financial Decision Support Systems, High Performance Computing, Logistics, Inventory, Supply Chain Management, Operations Research, Parallel and Distributed Computing, Performance Evaluation of Computer Systems, Project Management/ Information Flows, Telecommunication Systems, etc. Without simulation the modern technological world would fall apart. In the future we will depend on simulation more and more to predict the results of our actions and our inventions.

References

[1] *Modeling and Simulation Resource Directory*. The Society for Modeling and Simulation International. 7:1-17, ISBN 1-56555-2253-9, 2002.

[2] M. Shah. *Engineering Simulation: Tools and Applications Using the IBM PC Family*. Prentice Hall, ISBN 0-13-279183-8, 1988.

[3] R.G.E. Franks. *Mathematical Modeling in Chemical Engineering.* John Wiley, 1967.

[4] J. Banks et al. *Introduction to Siman and Cinema V.* John Wiley, ISBN 0-471-30960-5, 1995.

[5] G. Korn et al. *Digital Continuous System Simulation.* Prentice Hall, ISBN 0-13-212274-x, 1978.

[6] G. Korn. *Interactive Dynamic System Simulation.* McGraw-Hill, ISBN 0-07-852262-5, 1988.

[7] F. Cellier. *Continuous System Modeling.* Springer-Verlag, ISBN 0-387-97502-0, 1991.

[8] A. Pritsker. *Simulation and SLAM II.* Halsted Press, ISBN 0-470-20087-1, 1984.

[9] Y. Chu. *Digital Simulation of Continuous Systems.* McGraw-Hill, 1969.

[10] F. Speckhart et al. *CSMP-The Continuous System Modeling Program.* Prentice-Hall, ISBN 0-13-371377-6, 1976.

[11] T. Schriber. *Simulation Using GPSS.* John Wiley, ISBN 0-471-76310-1, 1974.

[12] J. Banks et al. *Discrete-Event System Simulation.* Prentice-Hall, ISBN 0-13-088702-1, 2001.

[13] F. Cellier (ed.). *Progress in Modeling and Simulation.* Academic Press, ISBN 0-12-164780-3, 1982.

[14] G. Bekey and W. Karplus. *Hybrid Computation.* John Wiley, 1968.

[15] N. McClamroch. *State Models of Dynamic Systems.* Springer-Verlag, ISBN 0-387-90490-5, 1980.

[16] Z. Karian et al. *Modern Statistical Systems, and GPSS Simulation.* W.H. Freeman and Co., ISBN 0-7167-8232-4, 1991.

[17] W. Graybeal and U. Pooch. *Simulation Principles and Methods,* Winthrop Publishers. ISBN 0-87626-811-4, 1980.

[18] C.D. Pegden et al. *Introduction to Simulation using SIMAN.* McGraw-Hill, ISBN 0-07-049217-4, 1990.

[19] W. Bulgren. *Discrete System Simulation.* Prentice-Hall, ISBN 0-13-215764-0, 1982.

[20] P. Bobillier et al. *Simulation with GPSS and GPSS-V.* Prentice-Hall, ISBN 0-13-810549-9, 1976.

[21] K. Dunning. *Getting Started in GPSS.* Engineering Press, ISBN 0-910554-34-x, 1981.

[22] A.W. Bennett. *Introduction to Computer Simulation,* West Publishing. ISBN 0-8299-0017-9, 1974.

[23] N. Karaynakis. *Computer-Assisted Simulation of Dynamic Systems with Block Diagram Languages.* CRC Press, ISBN 0-8493-8971-2, 1993.

PART III

Applications of Simulation

13

Parallel Simulation for Environmental Management

W.K. Giloi, T. Lux, M. Schmidt, A. Sydow and S. Unger
Fraunhofer-Institute for Computer Architecture and Software Technology (FIRST), Berlin, Germany
achim.sydow@.rst.fhg.de

Summary. This paper gives an overview of the research at Fraunhofer FIRST in the field of parallel simulation with application to environmental management. It presents the conception and history of the DYMOS model system and the interconnections between DYMOS and the further, more applied projects. The developed systems have been applied for winter smog, summer smog, ozone forecast, transboundary air pollution, traffic induced air pollution, environmental risk management and will be enlarged to include climate change and sustainable development in the near future. The general aim of DYMOS is to provide users with branch-oriented simulation systems and the know-how necessary for applying these systems.

1. Introduction

The history of computer simulation began with model descriptions based on ordinary differential equations solved on analog computers. The breakthrough of simulation applications in industry, research and management was achieved by the extension of model languages as input languages which are specific to user classes. At first, the simulation software consisted of microscopic basic elements from which the user had to form models/submodels. Later on, models/submodels were already included within the simulation system or could be easily involved using a suitable simulation framework. This development has been characterized by the enlargement from the analysis of large homogeneous systems to complex inhomogeneous systems. In this way, simulations have become very user-friendly. The acceptance is high because the used models as a whole are validated. In practice, the application is a compromise between run-time performance and convenience for users. Such branch-oriented simulation systems provide a maximum of software technology for the analysis of systems.

The DYMOS system [17] as a branch-oriented simulation system consists of models for air pollution analyses such as traffic flow and traffic emission models; microscale, mesoscale and macroscale atmospheric models; and chemical transport models. Thus the simulation system does not use only the mathematical or systems theoretical basic elements. It uses mainly clusters of these elements which form the models/submodels. The evolution of DYMOS as a simulation tool box for air pollution analysis is a result of selections of different models suitable for the configuration to a specific application.

Section 2 describes in detail the evolution of the DYMOS system as an exemplary branch-oriented simulation system. Section 3 lists special system implementations based on DYMOS and their usage for applications. Section 4 presents approaches to parallelize simulation models and results for the runtime behavior of parallel implemented models. Section 5 gives an outlook for further research directions.

2. The DYMOS History

In 1992 the DYMOS project started at Fraunhofer FIRST (at that time GMD FIRST) with its first objective being to analyse the dispersion of SO2 in the Berlin region. Gaussian models were used for this initial investigations.

Gaussian models were developed to study air pollutant dispersion in the neighborhood of big industrial facilities. They are based on an analytical solution of the advection-diffusion equation (transport equation). Gaussian models assume the homogeneity and stationarity of the wind fields, and because this case is exceptional in the atmosphere, their usage is strongly limited.

Since the model domain of DYMOS was the whole Berlin region, Gaussian models were not able to produce the required results. For this reason, the mesoscale Eulerian model REWIMET [6] was chosen for use in further investigations.

Eulerian models compute meteorological fields and dispersion processes on the basis of some balance equations (impulse derived from the *Navier-Stokes* equation, mass, energy, etc.) for model domains consisting of grid boxes. This leads to a system of partial differential equations which is approximated and parameterized in different ways, resulting in models of different complexity.

The parameterization of Eulerian models depends on the scale of the model domain to be considered. Historically, a division into 3 scales has been established: macroscale, mesoscale, and microscale. The macroscale normally deals with a range of 500 to 10000 km which includes phenomena like high and low pressure regions. Typical applications are air pollution simulations for the whole of Europe or the United States. The mesoscale deals with a range of 20 to 500 km which includes phenomena like tropical cyclones, sea breeze flows, urban heat islands, mountain lee waves, local storms, etc. Typical applications are urban and industrial regions, metropolitan areas, coastal regions, etc. The microscale describes the atmospheric flow processes below 20 km wavelength which includes local turbulences. Typical applications are street-canyon models, flows around buildings, identification of special polluters, etc.

The mesoscale model REWIMET is based on a hydrostatic, non-divergent and dry atmosphere. In contrast to true 3D Eulerian models which calculate the variables at the nodes of a locally fixed spatial grid, REWIMET only uses the fixed grid structure horizontally. Vertically, the model is subdivided into 3 layers lying on top of each other. A part of the model variables, namely the horizontal wind components, the potential temperature, and the air pollutant concentrations, is calculated for each horizontal grid point as box average in all 3 layers. The vertical wind component, the pressure, and the turbulent flux of impulse, heat and air pollutants are determined at the boundaries between the layers. The model REWIMET is driven by the suprascale stratification, the suprascale horizontal pressure gradient (geostrophic wind), and the surface temperature. The input of the geostrophic wind and surface temperature can be time-dependent. REWIMET considers the orography and the land use in the model domain. The transport of several air pollutants can be calculated simultaneously.

In order to be able to use a higher vertical resolution of the model domain and to include cloud physics where appropriate, the Eulerian model GESIMA [8] was added to the DYMOS system. GESIMA is a true three-dimensional, non-hydrostatic mesoscale model which realizes the inelastic *Boussinesq* approximations (density variations are considered only in the buoyancy term) and which includes cloud physics. The viscosity properties of air are neglected.

Due to new tasks evolving in further, more applied projects of the "DYMOS family", a Lagrangian model was included in the DYMOS system [3].

Lagrangian models also differ depending on the model scale. For microscale and mesoscale domains the transport of a set of representative particles is commonly modeled by an advective part (deterministic calculation), and a part of turbulent diffusion (stochastic calculation, e.g. by a *Markov* process). An exemplary macroscale model [15] only considers air columns moving by the advective wind. Emissions, sedimentations, and depositions are determined from the position of the particles or air columns resp. by integration. Due to their properties, Lagrangian models are especially suited for calculating air pollutants trajectories (e.g. transboundary analysis) and for localizing emission sources.

One major difference between Eulerian and Lagrangian models is that Eulerian models calculate meteorology and transport variables for the model grid, whereas Lagrangian models only calculate particle or air column transport. The meteorological fields are input variables for Lagrangian models.

To simulate the dispersion of primary species such as winter smog components (SO_2, dust, etc.) or substances of special interest (benzol, lead, etc.) only emission models and a meteorology/transport model is necessary. The dispersion simulation of secondary species like ozone and other photochemical oxidants additionally requires chemical transport models.

A chemical transport model is characterized by the chemical kinetic mechanism to describe the rates of atmospheric reactions including homogenous gas-phase, heterogeneous, and liquid phase reactions. At present, comprehensive models explicitly treat a wide variety of sulfur compounds, nitrogen compounds, and hydrocarbon species, along with the important oxidizing species such as O_3, OH, and H_2O_2.

The chemistry rate equations of the mechanism comprise a system of ordinary nonlinear differential equations. Because of the huge differences in reaction kinetics, such systems are extremely stiff and therefore the solution generally requires hours of computing time. The integration of the chemistry rate equations typically consumes 90 per cent of the total computing time of an air pollution simulation including air chemistry. Therefore the choice of the algorithm for the solution of the chemistry equations plays an important role.

Since 1993, the DYMOS project has been concerned with the analysis of summer smog in the Berlin region. For this reason, the chemical transport model CBM-IV [4] was added to the DYMOS system.

CBM-IV is a popular and sufficiently tested reaction scheme describing the most important chemical processes in gas phase chemistry for the production of ozone and other photooxidants. It is officially distributed by the Environmental Protection Agency of the United States. CBM-IV is a condensed version of the original CBM. Carbon atoms with similar bonding are treated similarly. There is no need for the definition of an average molar weight, so that this mechanism is mass balanced. Some species are handled explicitly because of their special character in the chemical system (for example isoprene which is the most commonly emitted biogenic species). The mechanism involves 34 species and 82 reactions, and contains 9 primary organic compounds. To profit from the features of the CBM-VI, detailed information on the hydrocarbon mixture is necessary.

A major precondition in order to perform air pollution simulations is the availability of detailed and, if possible, dynamic emission data. According to the group of polluters (industry, private households, traffic), the effort involved in pre-processing the available data into the input data for the chemical transport models differs considerably.

In case of industry emissions (power plants, metallurgy, etc.) there are generally point sources with fixed coordinates. The emission model here is only an algorithm to calculate the dynamic function of emission amounts per hour from the reported amount per year by means of data about the emission variation over a 24-hour period and a 7-day period, respectively.

The emissions of private households are summarized to area sources proportional to population density according to the shape of the model grid used. The procedure to get emission amounts per hour from amounts per year is the same as for industry.

The traffic emissions with a share of more than 70% of all emissions are mainly responsible for high concentrations of air pollutants, especially when different substances of the exhaust gas react to ozone during hot summer days. The exact description of traffic emissions in urban areas is a time consuming process. Traffic emissions are influenced by a number of parameters, including internal parameters like engine size, engine type, different kinds of catalysts or exhaust gas emission reduction devices, the engine temperature as well as the determination of the traffic volume and the driving parameters of a single vehicle. To model traffic emission the following main tasks are to be performed:

1. vehicle type analysis

2. determination of exhaust emission factors
3. analysis of the driving parameters of the vehicles
4. determination of the traffic volume
5. registration of the geometric features of the road network.

In order to provide dynamic traffic emission data, steps 3 and 4 must be performed by a traffic flow model. Traffic flow models are an approximate description of specific phenomena within the traffic flow operations, normally based on a set of mathematical equations or on an analogy to other physical systems [9].

There are two basic concepts in traffic flow simulation. The first covers microscopic models in which the movement of individual vehicles is simulated. The computational effort for these models increases rapidly with the number of vehicles considered.

The second class consists of macroscopic models in which aggregate variables such as traffic density, mean speed and volume describe the traffic flow. Due to the representation by aggregate variables, less computation time is required than with microscopic variables. On the other hand, macroscopic models are not able to provide information on travel time, fuel consumption or route choice, factors which would be of particular relevance when the macroscopic model is coupled with an air pollution simulation system. The model DYNEMO [14], selected for usage together with DYMOS, combines the advantages of both the microscopic and macroscopic traffic-flow models.

DYNEMO is a simulation tool for both urban and rural road networks. It has already been used to simulate large parts of the German motorway network. Regarding the movement of vehicles, DYNEMO is a mesoscopic model in the sense that it combines properties of both microscopic and macroscopic models. As in microscopic models, the unit of traffic flow is the individual vehicle. Their movement, however, follows the approach of macroscopic models and is governed by the average traffic density on the link the vehicles travel.

Integrating these models described above (see Figure 1), the DYMOS system is able to support users in governmental administrations and industry in operative decision making as well as long-term planning, to serve as a scientific basis for national or international negotiations, and to enable predictions (air pollutants concentration, traffic density, etc.) for informing the public via WWW (see Figure 2), TV or radio.

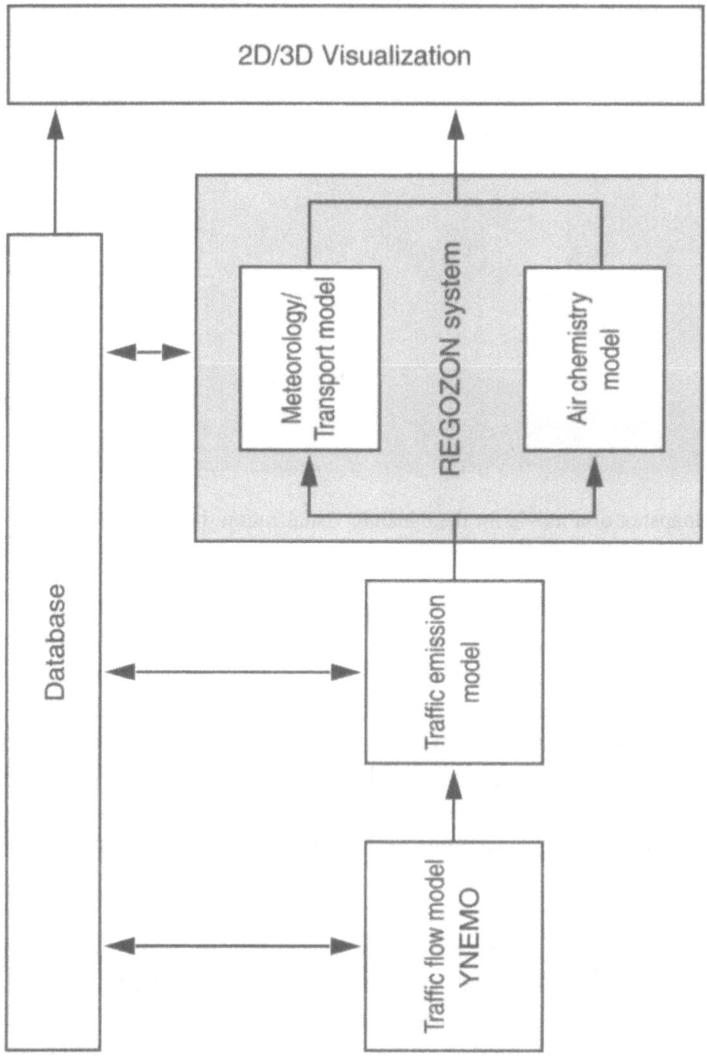

Figure 1. DYMOS System Structure

Today the DYMOS project is faced with two additional challenges: global change and sustainable development. For the simulation of climate variability, general circulation models (GCM) are used. A new approach in this field [19] is characterized by the coupling of an atmospheric GCM with an oceanic GCM.

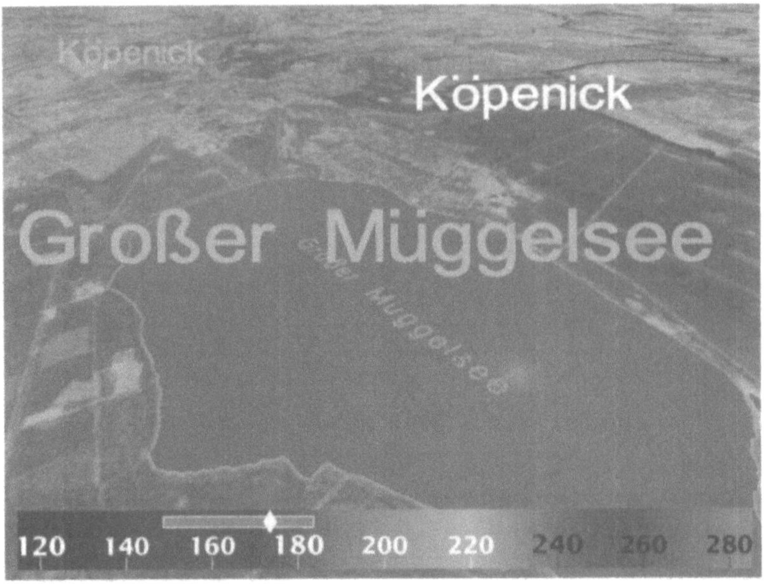

Figure 2. Snapshot of a movie for the dynamic visualization (helicopter flight) of near surface ozone concentrations in the Berlin region

Since the 1970s, there has been a growing interest in an integrated approach to the problem of global change and sustainable development, called integrated assessment. In general, integrated assessment can be defined as an interdisciplinary process of combining, interpreting and communicating knowledge from diverse scientific disciplines in such a way that the whole cause-effect chain of a problem can be evaluated from a synoptic perspective with two characteristics: it should have added value compared to single disciplinary oriented assessment and it should provide useful information to decision-makers.

In Europe, integrated assessment has its origins in population environment, ecological and acidification research. In North America, attention was mainly concentrated on economic modeling. During the past few years, integrated assessment models have been increasingly focused on climate change and sustainable development.

The main questions under consideration today are: human health management and population growth, management of fossil fuels and renewable energy resources, safeguarding of food and fresh water and the stability of the biogeochemical cycles with respect to human perturbations.

To model these complex systems, the Pressure-State-Impact-Response concept may be used as organizing principle for achieving a plausible division of the cause-effect chains into subsystems. The Pressure System represents social, economic and ecological driving forces underlying the pressure on the human and environmental system. The State System represents physical, chemical and biological changes in the state of the biosphere, as well as changes in human population and resources/capitals. The Impact System represents social, economic and ecological impacts as a result of human and/ or natural disturbance. The Response System represents human intervention in response to ecological and societal impacts.

Modern integrated assessment approaches use a hierarchical structure to overcome the complexity problem of modeling. At the lower level of aggregation, economy-energy models operate in multi-year time steps with national or regional political boundaries. Their regional and distributed output data may be used in theme-specific models like RAINS (acidification) or IMAGE (climate) at the next level of aggregation. Their results of characteristic simulations provide a guide line to setting up metamodels which establish the building blocks for integrated assessment models at the highest level of aggregation.

3. System Development and Applications

The first software implementation within the DYMOS project was the REGOZON system. The REGOZON system consists of the mesoscale Eulerian model REWIMET and the chemical transport model CBM-IV. Various case studies of summer smog conditions in urban areas have been carried out using the REGOZON system [10]. The Department of the Environment of the Berlin state government and the Ministry for the Environment of the state Brandenburg commissioned summer smog analyses for the results of the FLUMOB measuring campaign carried out in July 1994. Greenpeace commissioned an analysis of the influence of emissions caused by traffic in Munich on the ozone concentration in the Munich area. The analysis was performed for a typical mid-summer day in 1994. Within a DYMOS subproject, the REGOZON system coupled with a Petri-net-based traffic flow model was used to analyze the traffic induced air pollution of the city of Budapest.

In order to consider different environmental domains such as air, surface water and coastal water and therefore to ensure as complete an environment quality analysis as possible, the European cooperation project ECOSIM [11] was initiated by the DYMOS group.

The ECOSIM system builds on a set of sophisticated state-of-the-art simulation models (e.g. air chemistry models). It exploits existing knowledge in environmental management by connecting these models with an advanced GIS, on-line monitoring networks, local databases, and an embedded expert system. This allows e.g. a direct link between the models, up to-date measurements of current meteorological conditions, and actual pollutant background concentrations. The following simulation models are currently implemented into the ECOSIM system:

1. the non-hydrostatic, prognostic, grid-based (Eulerian) mesoscale 3D windfield model MEMO [12];
2. the dispersion and air chemistry part of the DYMOS system, responsible for the advective transport, turbulent diffusion and surface uptake of the air pollutants;
3. the mesoscale ozone forecast model REGOZON;
4. the modular, finite-difference based 3D ground water model MODFLOW-96;
5. the prognostic 3D ocean model POM, responsible for the temperature-, velocity-, and salinity distribution in shallow water as well as in deep ocean domains.

The ECOSIM system integrates data acquisition and monitoring systems, GIS and dynamic simulation models in a flexible client-server architecture. In general, the different model components run at different locations on different computers. They are connected via Internet and controlled by a main server located at the user's office. The simulation results can be visualized on the server as well as on any remote client's platform.

ECOSIM has been developed and tested in close collaboration with three city authorities with differing environmental problems and state-of-the-art approaches to environmental management. ECOSIM was used within the city of Berlin, Germany, to provide an improved method for modeling the level of ozone which depends mainly on traffic emissions. For the city of Athens, Greece, similar problems have been addressed as well as the interplay between surface and ground water and coastal pollution. ECOSIM was also applied in a more limited manner in the city of Gdansk, Poland.

The aim of the European cooperation project HITERM [18] was to reach better-than real time performance for the simulation of the accidental release of hazardous substances into the atmosphere, ground, or surface water. The developed HITERM system will support the user in the field of emergency management, emergency planning, and emergency training, both for transportation accidents and (Seveso-class) chemical process and storage plant accidents. Integrating environmental risk criteria and road information, the system will be able to simulate accident scenarios along the preferred and a number of alternative routes, and to determine the optimal routing (dynamic adaptive routing).

In the HITERM project, the DYMOS group was developing an air quality modeling system with special emphasize to manage accidents with toxic substances. For this purpose, the model system must run in a time much shorter than real time, usually with a sparse input data set. This requires a careful design of all model compounds and its implementation. The model contains air pollutant release models, a diagnostic wind field model and a Lagrangian particle model. The model systems includes uncertainty analysis tools to take into account the serious deficits with regard to input data quality.

Among the performed case studies to test and demonstrate the capacities of the system are:

1. dynamic adaptive routing (Portugal): transportation of hazardous cargo (petrol and diesel) by a vehicle fleet;
2. transportation accident simulation (Gotthard alpine corridor, Switzerland): release of a hazardous substance into the atmosphere or into the ground water;
3. process plant accident (Region of Ponte S. Pietro, Northern Italy): leaking or spilling of toxic or flammable substances into the environment.

The objective of the SIMTRAP project [13] was the development and demonstration of an integrated simulation system for traffic flow, traffic emissions and air pollution dispersion as a decision support tool for transport and environmental authorities.

SIMTRAP centers around two well-established core components: the mesoscopic dynamic traffic model DYNEMO and the air pollution model system DYMOS. Due to the complexity of the simulated processes, both modules are being implemented in a remote HPCN environment. The simulation results will be visualized in a local 3D GIS system with built-in functionality for decision support. Communication is realized using existing computer networks and standard protocols.

The SIMTRAP prototype will be applied by potential users at four test sites to demonstrate the functionality of the system, and to validate the results gained. The test applications vary in size and objective from urban street networks in Berlin and Vienna, via a regional network in the Milan area, to an interurban highway network in the Ranstad area in the Netherlands.

Within the SIMTRAP project, the DYMOS group was especially concentrating on the parallelization of the dynamic traffic flow model DYNEMO in order to receive information about the temporally and spatially distributed traffic induced primary pollutants. A special multicriterial decomposition technique had to be developed, emphasizing equal distributed processor workloads and minimal communication effort. Another main focus was the implementation of a new emission model taking into account all the microscopic driving information of the individual cars from the traffic flow model as well as the cold start emission behavior and the road type. SAS was also responsible for adapting the complete SIMTRAP prototype for the Berlin/Brandenburg area.

Another focal point of the DYMOS group are European cooperation activities in the field of long-range transport of air pollutants and the environmental effects in different areas of Europe in order to assess alternative control strategies. Main partners here are the International Institute for Applied Systems Analysis (IIASA), Laxenburg (Austria) and the Norwegian Institute for Meteorology, Oslo (especially the EMEP's Meteorological Synthesizing Center West - MSC-W). In the subproject DYMEMEP [16] the Lagrangian model of the EMEP/MSC-W for calculating the transboundary transport of air pollution in Europe has been parallelized and implemented on the Cray T3E computer of the Norwegian University of Science and Technology in Trondheim.

4. Model Parallelization

Air pollution simulations (or environmental applications in general) require extremely large amounts of computing time and memory. In order to make the results of case studies available to users within an acceptable period of time or to enable a smog prediction to be made at all (computing time less than simulation period), it is necessary to parallelize environmental simultion models. In this section exemplary approaches are described for parallelizing different models.

4.1 EMEP's Lagrangian Model

The parallelization of this model became necessary to perform the large number of scenario runs necessary to derive the input data for IIASA's (International Institute for Applied Systems Analysis, Laxenburg, Austria) Integrated Assessment Model [7]. This model contains a regressor model, which allows to predict consequences of emission scenarios by applying a simple formula. Coefficients of this regressor equation were derived by means of the data computed with the Lagrangian model. This results in the necessity to perform 1200 scenario runs, which had been impossible on a sequential computer, but could be done on 8 processors of the T3E in Trondheim in about a week. The parallel version of the program was used for the extensive regular calculations at EMEP, too.

The main problems in parallelization of the Lagrangian model are the following:

1. Initial concentrations depend on concentrations computed in previous time steps.
2. Trajectories may have different lengths; hence computational work per trajectory may be quite different.

The first problem restricts the granularity of parallelization to a single time step - this does not cause large problems, since the CPU-time per time step remains in the range of several seconds, which is long enough to gain performance from parallelization. Second problem leads to difficulties with the load balancing.

The parallelization was done using the master/slave architecture: One node does all the input/output work and acts as a supervisor of the whole calculation. It also computes the trajectories, distributes the work between the other nodes and aggregates the results. The other nodes (slaves) act as calculators of the concentrations on the trajectories, which is the main work to be done.

The distribution of the work between the nodes is done in a simple but effective way: The trajectories are ordered according to their length, and then each node gets an almost equal number of trajectories of the same length for the calculation. This ordering has the following effect for the organization of the computations: In the original program almost all do-loops began with a statement like 'if the trajectory already exists at the given time step, then perform the following calculations', which did not give the compiler much possibilities for optimization. Having an ordered array of trajectories (where the first one is the longest) one can simply replace all these do-loops by loops with an upper bound, giving the actual number of 'active' trajectories at a given inner time step of the calculation. The if-branches can be removed and the compiler has a lot of possibilities for optimization.

This leads to a speed up of 2 for the serial program in a base case run. For a scenario run this speed up increases, since the unchanged trajectories now do not take part in the computations at all, whereas in the original version the if-tests had to be done for all trajectories. Consequently, whereas a scenario run with the original model needed about 40-50% of the CPU time of a base case run (for 15-30% of work), now the CPU time fraction is nearly equal to this fraction of the necessary work.

The above mentioned first problem turns out to be not so crucial, since only a very small amount of trajectories needs initial values from the previous time step (most use values from several days before). These trajectories can be computed on the master node, allowing an overlay of input/output, communication and calculations leading to a nearly continuous work of the slaves.

Figure 3 shows the speed-up and parallel efficiency (defined as CPU time on n nodes / CPU time on a single node and speed-up / number of nodes, correspondingly) for the calculation of one month real time on 1 to 8 processors. This speed up is highly super-linear up to 32 nodes, reflecting the fact of better cache performance in case of smaller concentration arrays.

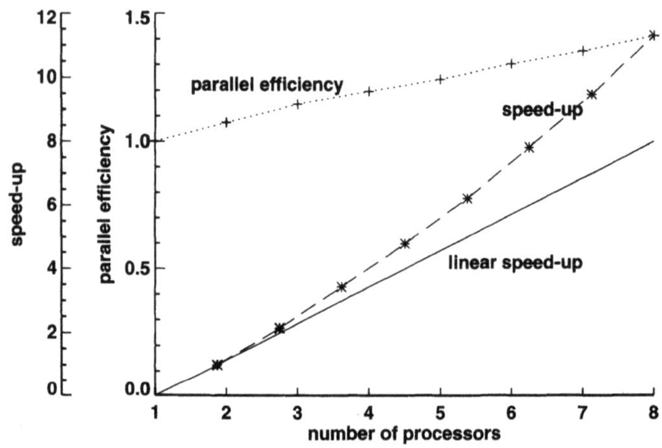

Fig 3. Speed up and parallel efficiency for the only ordered version of the Lagrangian model

4.2 EMEP's Eulerian Model

These programs are parallelized by dividing the horizontal domain in rectangular boxes. Beside the input/output operations, communication is only necessary in the calculation of horizontal advection, which consists of subsequent Bott steps in each of the horizontal directions. For the concentrations 3 halo values are necessary to be taken from the neighboring processors, while for the velocity components 2 values suffice. Consequently, the communication structure is as follows: exchange halo values in one horizontal direction, do advection calculation in this direction, exchange halo values in the other horizontal direction, do advection calculations in this direction. There is some small overhead introduced in the calculations of the coefficients for Bott's advection scheme, since there are inner boundaries introduced on the distributed blocks. Main reason for not reaching linear speed-up are load balancing problems, since the actual dimensions of the computational domain do not t exactly the distribution, resulting in differences in the number of points per processor. Nevertheless, the speed-up is nearly linear and for some parts of the models even super-linear.

4.3 The DYNEMO Model

The simulation of large traffic networks with substantial numbers of cars requires considerable run times. The traffic flow model DYNEMO has been parallelized in order to speed up this simulation process. One of the major problems is to develop a suitable communication structure of the parallel processes. First results have shown that an ordinary master slave topology appears to be well suited.

The master process is responsible for supervising the slaves, the initialization and decomposition of the traffic network as well as for such tasks requiring information of the state of the complete network, e.g. the dynamic route choice. During one time step, each slave calculates the spatial motion of the vehicles situated on their sub-networks. The slaves are arranged on a nonregular 2D-grid, whereas the shape of this 2D-grid depends on the decomposition of the network. In fact, the decomposition has to satisfy partly contradictory aims. On the one hand, the decomposition should provide possibly large coherent regions of the sub-networks with minimal interconnections (stretches) between the sub-networks. The number of stretches influences the number of cars moving to a neighboring sub-network. Since car exchange between sub-networks requires time for interprocessor communication, this approach guarantees a minimal communication effort.

On the other hand, the decomposition has to ensure a nearly equally distributed workload on all physical processors. This workload depends not only on the number of stretches or lanes but also on the number of vehicles driving on the sub-network. Unfortunately, the traffic on the roads is a priori unknown. Assuming an equally distributed traffic density on all stretches, the following approximation of the workload ***Wn*** seems to be reasonable

$$W_n = \sum_{i=1}^{L_n} section(i).lane(i), \qquad \ldots(1)$$

where *section(i)* denotes the number of sections of the i-th stretch, *lane(i)* the number of lanes of the i-th stretch and *Ln* the number of stretches of the nth sub-network.

Equation (1) has been applied to a network of the inner city of Berlin. Since traffic changes rapidly during the day, a dynamic load balancing may be required. In case of a dynamic load balancing, a rearrangement of the topology of the communication structure has to be performed from time to time. Since this procedure is very time consuming, we only have set up this option at the moment. At a later time we will decide whether to implement the dynamic load balancing depending on the run-time results or not. The traffic flow model has been implemented by means of the PVM [2] message passing model.

The Model Server can be implemented on a dedicated parallel machine (at Fraunhofer FIRST within this project, but ultimately at any convenient site), or on a UNIX workstation cluster or multi-processor system (taking into consideration the description of hardware requirements).

Straightforward protocols are defined for passing a simulation request from the SIMTRAP Server to the Model Server, and for passing back the results.

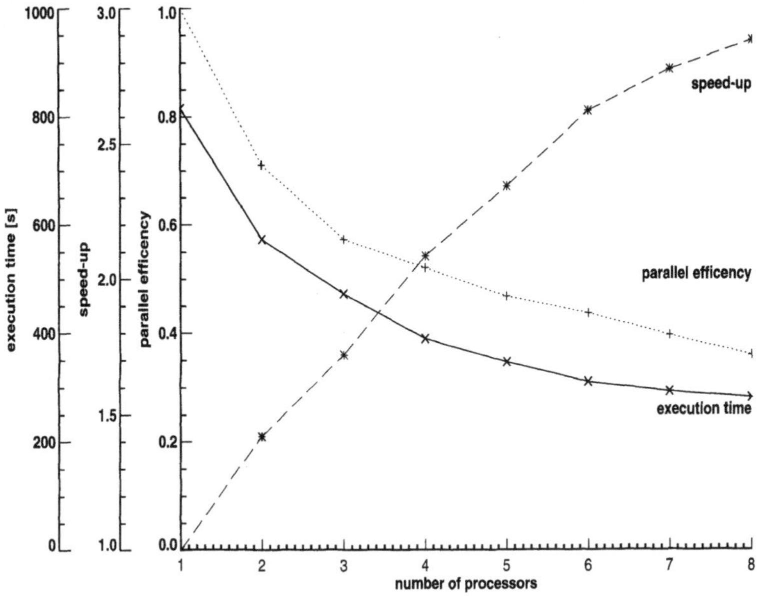

Figure 4. Execution time, speed-up and parallel efficiency for 600 time steps (10minutes) of the parallelized DYNEMO version.

Figure 4 displays the execution time, speed-up and parallel efficiency as a function of the number of processors. The simulation has been performed on the traffic network Berlin-Brandenburg as described above. The underlying hardware platform was a 8 node 266 MHz Pentium2 system with each 128 MByte memory, connected by a fast Ethernet (100 MBit) running under LINUX. The results have been received by simulating 600 time steps corresponding to 10 minutes real time. The speed-up scales sufficiently up to 8 processors with a maximal run time about twice as fast as real time. Because of the unequal load balancing, a further increase of the number of nodes appears not meaningful at the moment. In order to improve the load balancing we will introduce a new decomposition criteria taking into account a more realistic traffic load from the origin-destination matrix.

5. Outlook

An important point for further research is the design of simulation frameworks endowed with virtual world capabilities establishing a new quality in simulation of complex socio-economic and environmental problems. The virtual world concept supports a spatial metaphor for even complicated problems, high-end visualization and a deep experience with distributed simulation environments. By interaction of agents (human and artificial) within a unique environment, the special qualities of human and computer-based information processing are integrated. This will open a new view and potentials for the investigation of sustainable developments.

At Fraunhofer FIRST this concept is realized within the Man-Model-Measurement (M3) system [1], which differs from traditional approaches in its intuitive presentation of simulation results and its involvement of real-life actors in the simulation. These human actors are able to act in a virtual world, as created on the basis of the multidisciplinary simulation models. This enables them, in turn, to influence this world and the models' input data. The M3 simulation thus combines the scientific approach of conventional simulation systems with the multimedia possibilities of Virtual Reality (VR).

Besides studies on sustainable development, future work is likely to focus on traffic telematics and control as well as grid-based environmental analyses which can be used to minimize risks in the case of accidents and malfunctions. In the area of grid computing, software tools to rapidly implement distributed applications on the Fraunhofer Resource Grid (FhRG) are under development.

References

[1] T. Asselmeyer-Maluga, P. Frank, A. Hoheisel, H. Rosé, B. Walter. The M3-Project: Distributed Environmental Simulation and Virtual Reality. *In: Simulation in Industry / Giambiasi*, Norbert [Ed.], 573 - 577, 2001.

[2] A. Geist, A. Beguelin, J. Dongarra, W. Jiang, R. Manchek and V. Sunderam. *PVM: Parallel Virtual Machine - A Users' Guide andTutorial for Networked Parallel Computing*, 1994.

[3] I. Gerharz, A. Sydow, S. Unger. "Determination of Sources of Particles in Air by Trajectory Models". *Proc. 2nd IMACS Symposium on Mathematical Modeling (MATHMOD'97)*, Vienna, Austria, February 5 - 7, 1997.

[4] M.W. Gery, G.Z. Whitten and J.P. Killus. "Development and Testing of the CBM-IV for Urban and Regional Modeling". US Environmental Protection Agency, EPA-600/3-88-012, USA, 1988.

[5] M.L. Jugel, A. Sydow. "Parallelity in High Level Simulation". *Transactions of SCS*, 15 (3): 101-103, 1998.

[6] D. Heimann. "Ein Dreischichten-Modell zur Berechnung mesoskaliger Wind- und Immissionsfelder über komplexem Gelände". Ph.D. thesis, University of Munich, Germany, 1985.

[7] C. Heyes, W. Schöpp, M. Amann and S. Unger. "A Simplified Model to Predict Long-Term Ozone Concentrations in Europe". WP-96-12, *International Institute for Applied Systems Analysis (IIASA)*, Laxenburg, Austria, 1996.

[8] H. Kapitza and D.P. Eppel. "The Non-Hydrostatic Mesoscale Model GESIMA. Part I: Dynamic Equations and Tests". *Beitr. Phys. Atmosph.*, 65: 129-146, 1992.

[9] H. Lehmann, B. Kwella. "Empirical Traffic Flow Modeling: "Floating Cars" in Urban Networks", *Transactions of SCS*, 15 (3): 133-136, 1998.

[10] P. Mieth, S. Unger. "Comparison of Identical Emission Reduction Measures under Different Meteorological Conditions". *Proc. 4th International Conference on Air Pollution*, Toulouse, France, August 28 - 30, 1996.

[11] P. Mieth, S. Unger, M. Jugel. "An Environment Simulation and Monitoring System for Urban Areas", *Transactions of SCS*, 15(3):115-121, 1998.

[12] N. Moussiopoulos, T. Flassak, D. Berlowitz, P. Sahm. "Simulations of the Wind Field in Athens with the Nonhydrostatic Mesoscale Model MEMO". *Environmental Software*, 8: 29-42, 1993.

[13] M. Schmidt, K. Nökel, R.-P. Schäfer. "SIMTRAP - Simulation of Traffic Induced Air Pollution". *Transactions of SCSI*, 15(3): 122-132, 1998.

[14] T. Schwerdtfeger. "DYNEMO: A Model for the Simulation of Traffic Flow in Motorway Networks". *Proc. 9th Int. Symposium on Transportation and Traffic Theory*, 1984.

[15] D. Simpson. "Long Period Modeling of Photochemical Oxidants in Europe". *Atmos. Environ.*, 26A: 1609-1634, 1992.

[16] D. Simpson, K. Olendrzynski, A. Semb, E. Storen and S. Unger. "Photochemical Oxidant Modeling in Europe: Multi-Annual

Modeling and Source-Receptor Relationships". EMEP/MSC-W Report 3/97, 1997.

[17] A. Sydow, T. Lux, P. Mieth, S. Unger. "The DYMOS Simulation System for Air Pollution Analysis". *Proc. SCS 1997 Western Multiconference*, Phoenix, Arizona, January 12 - 15, 1997.

[18] P. Mieth, S. Unger, I. Gerharz. "A Model-based Tool for Environmental Risk Management after Accidental Release of Toxic Substances". *Systems Analysis Modeling Simulation,* 41: 325-337, 2001.

[19] J.-Y. Yu, C.R. Mechoso. "A Coupled Atmosphere-ocean GCM Study of the ENSO Cycle". *Journal of Climate*, 14: 2329-2350, 2001.

14

Role of Intracellular Ca^{2+} Dynamics in Supporting Spiral Wave Propagation

[1]B.Y. Kogan,
Departments of Computer Science

S. Lamp and J. Weiss,
Medicine (Cardiology), Physiology and Cardiovascular Research Laboratory, and The David Geffen School of Medicine at UCLA,

University of California, Los Angeles, CA 90095-1679
kogan@cs.ucla.edu

Summary. The intracellular Ca^{2+} dynamics play a key role in the transition from ventricular tachycardia to ventricular fibrillation through their influence on Ca-sensitive membrane currents. That was shown in our previous publication which deals with modified Luo and Rudy cardiac cell model. Here using the same model we study the effect of increasing sensitivity of non-specific calcium current, $I_{ns(Ca)}$, to $[Ca^{2+}]_i$ on wave propagation in two dimensional tissue model under conditions of ventricular tachycardia. Under these conditions some of the tissue cells come across high rates of stimulation with subsequent calcium overload and release from the Junctional Sarcoplasmic Reticulum.

Our results show that increasing sensitivity of $I_{ns(Ca)}$ to $[Ca]_i$ leads to progressive prolongation of the action potential duration in some of the tissue cells. That causes a non-uniform increase in the length of the propagated spiral wave to such an extent that the size of the tissue becomes insufficient for further propagation. The temporary block of propagation causes the appearance of cells with single and multiple *early after depolarization* (EAD), which formed in a portion of space with retarding repolarization. This area is the source of consequent initiation of direct or backward spiral waves, when the surrounding tissue goes out of excitation.

Keywords: Calcium dynamics, delayed after depolarization, instability of spiral wave, spontaneous Ca release, parallel computer simulation

1. Introduction

Most of the computer simulation studies of wave propagation in two and three dimensional (2D and 3D) cardiac tissue models were accomplished using cell models of the first generation [1, 2, 3] and their simplified versions

[1] Corresponding author.

[4, 5, 6, 7]. These cell models do not reproduce the intracellular Ca^{2+} dynamics.

New achievements in the study of cell Ca^{2+} dynamics and the need to simulate the physiological phenomena related to excitation/contraction coupling processes in the heart, facilitate the creation of cell models of the second generation with various formulations of intracellular Ca^{2+} dynamics [8, 9, 10, 11, 12]. For the isolated cardiac cell model, Luo and Rudy in [11] showed that when the sarcoplasmic reticulum (SR) is overloaded (to the predetermined threshold) it produces a spontaneous Ca^{2+} release. Through activation of Ca^{2+} dependent currents I_{NaCa} and $I_{ns(Ca)}$, spontaneous release may cause phenomena such as delayed after depolarization, early after depolarization (EAD), or spontaneous rhythmic and triggered activities.

In our previous studies [13, 14] we showed that rapid pacing of cells in 2D cardiac tissue model (observed under conditions of tachycardia and fibrillation) leads to Ca^{2+} accumulation in SR and its overload with the subsequent Ca^{2+} spontaneous release. The latter increases the intracellular free $[Ca^{2+}]_i$ and through initiation of the inward currents $I_{ns(Ca)}$ and I_{NaCa}, causes the prolongation of the action potential duration (APD). Because this prolongation of APD is non-uniformly distributed in space, the smooth stationary spiral wave becomes deformed and that causes the transition to non-stationary propagation and in some cases to wave front breakup. These phenomena are observed even when the APD restitution curve of tissue cells originally has a slope less than unity but become more pronounced when the slope is above unity.

Using model [14], we observed new patterns of non-stationary spiral waves in 2D homogeneous tissue by varying the parameters $K_{mns(Ca)}$ during computer simulation studies. When these parameters have their original values, it is possible to observe, after some time of non-stationary propagation, that all the tissue cells are at different levels of repolarization phase and further propagation becomes impossible. Usually, this situation leads, after some time, to full recovery of the tissue from excitation. But in our case when the coefficient $K_{mns(Ca)}$ was equal to 0.9 μM (in comparison to usually used 1.2 μM) and after most of the tissue went out of excitation, the remainder showed some activity. Shortly, the new spiral wave front appears, rotating in the previous or in the backward directions. These wave fronts are usually initiated in close vicinity to the border of the tissue model. We hypothesize that this phenomenon is caused by the appearance of the action potential (AP) with EAD distributed non-uniformly in this part of tissue. The reciprocal of the coefficient $K_{mns(Ca)}$ plays the role of an amplification factor, which increases the effect of spontaneous Ca^{2+} release from SR on $I_{ns(Ca)}$.

In this paper we present the details of the computer simulation of this phenomenon on a parallel supercomputer, and the major results and explanations. The origin of this phenomenon is connected with the fundamental properties of the Ca^{2+} dynamics (Ca^{2+} accumulation and its spontaneous release from SR under high pacing rate) rather than with specific properties of the model being used. The Ca^{2+} accumulation and its subsequent spontaneous release were observed in differently formulated intracellular Ca^{2+} dynamics models including the original Luo-Rudy model [11] as well as in physiological experiments [15, 16]. We hope that the block of propagation and subsequent origination of spiral wave rotation in the forward and backward directions, which we observed in computer simulation, will sooner or later be revealed in direct physiological experiments.

2. Method

The excitation wave propagation was studied in a 2D model of isotropic cardiac tissue, considered as a syncitium. The mathematical model of 2D tissue used for computer simulations has the form:

$$V_t = D\,(V_{xx} + V_{yy}) + I_{memb} + I_{stim}(x, y, t) \qquad (1)$$

where V (mv) is transmembrane potential, D (cm^2/ms) is a diffusion coefficient, I_{stim} ($\mu A/\mu F$) is stimulating inward current, and I_{mem} ($\mu A/\mu F$) is total transmembrane current:

$$I_{memb} = I_{Na} + I_{K1} + I_{Ks} + I_{Kr} + I_{Kp} + I_{Ca(L)} + I_{NaCa} + I_{NaK} + I_{ns(Ca)} + I_{p(Ca)} + I_{Nab} + I_{Cab} \qquad (2)$$

The expression for each current in (2), the equations for Ca^{2+} dynamics and the set of original parameters were the same as those in the modified LR cell model [14].

To make computer simulation feasible the model [14] was implemented on a massively parallel supercomputer, CRAY T3E, using the splitting operator algorithm [17]. According to this algorithm the integration of (1) is split into two parts: integration of the diffusion equation $V_t = D(V_{xx} + V_{yy})$ and integration of the rest of the system, the ordinary differential equations (ODEs) $\frac{dV}{dt} = I_{memb} + I_{stim}$. These integrations are executed in consecutive time cycles of predetermined duration Δt. The specific features of the application of this algorithm to the solution of the problem (1) on a parallel computer are described in our previous publications [13, 18]. The splitting operator algorithm allows integrating the ODEs at any point in space independently and with variable time steps.

The domain of integration was chosen as a square grid of 256x256 nodes connected by diffusion (D = 1 cm^2/ s). The parallel processors are distributed over the entire grid in stripes, so that each processor integrates only in its own part of the grid. The communication between processors is required only for solution of the diffusion equation and is implemented using a Message-Passing Interface. For all computer simulations we used time steps varying between 0.005 and 0.1 ms and fixed space step equal to 0.025 cm. The velocity of rectilinear front propagation is 55 cm/s.

Methodologically, we divided all the studies into two parts: 1) the effect of the increasing sensitivity of the inward currents $I_{ns(Ca)}$ to $[Ca^{2+}]_i$ on the behavior of an isolated cardiac cell under normal conditions and under a high rate of stimulation with and without Ca^{2+} spontaneous release, 2) the effect of this increased sensitivity on wave propagation in 2D tissue on the presence of spiral waves. It is worth noting that during spiral wave propagation, some of the tissue cells enter conditions close to those observed in single cells under high pacing rate.

3. Results

3.1 The cardiac cell model [14] under normal pacing rate but with different coefficients $K_{mns(Ca)}$ in the expression for $I_{ns(Ca)}$ current.

The parameters of the AP, its shape, intracellular free calcium concentration $[Ca^{2+}]_i$ and sodium current I_{Na} (see Figure 1A) are the same for both values of $K_{mns(Ca)}$ =1.2 and 0.9 μM .

Insignificant differences were observed in the $I_{ns(Ca)}$ currents and APD restitution curves (Figure 1B). With $K_{mns(Ca)}$ =1.2 μM , the $I_{ns(Ca)}$ is smaller and a small region of the diastolic interval (DI) of the APD restitution curve with slope equal to 1 is slightly larger than for $K_{mns(Ca)}$ = 0.9 μM .

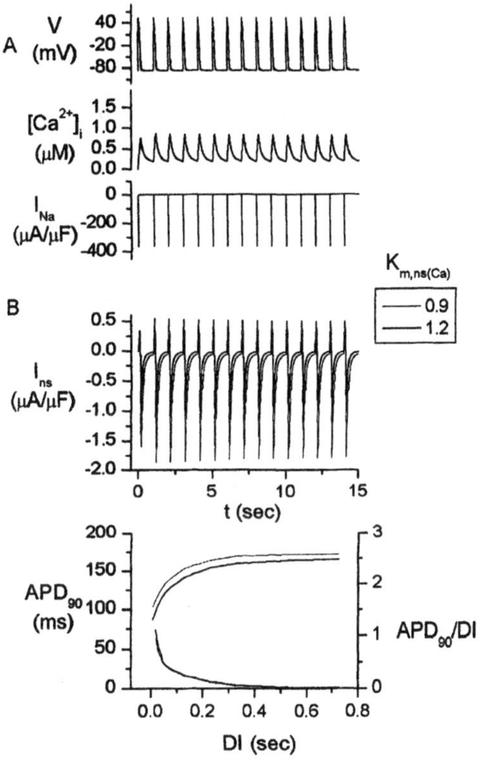

Figure 1. The cardiac cell model under normal pacing rate (BCL = 1000 ms) when $K_{m,ns(Ca)}$ = 0.9 and 1.2 μM

3.2 The cardiac cell model under high pacing rate but with different coefficients $K_{mns(Ca)}$

The pacing rate is chosen such that the basic cycle length (BCL)= 180 ms. That is close to the period of the spiral wave originating in 2D cardiac tissue composed of the cell models [14]. Because the pacing interval is smaller than the normal AP duration, a small transient period occurs in establishing the corresponding cell rhythm (see Figure 2 A, B, the first three beats).

From the very beginning of pacing, the accumulation of $[Ca^{2+}]_i$ is observed (Figure 2D). It is accompanied, after the first 1000 ms, with consequent appearance of spontaneous releases (Figure 2C). For each spontaneous release, the EAD occurs during the repolarization phase of the corresponding AP, which prolongs the overall duration of the AP. The depolarization rate here is determined basically by the stimulus current because the sodium current becomes negligibly small after the first 1000 msec of stimulation (Figure 2E). The voltage level at which EAD occurs, depends on the rebalance of the inward and outward currents due to spontaneous release. The distortion of current balance occurs during the repolarization phase, prior to the moment of the next beat application. It is worth noting that the period of spontaneous release is larger than the period of stimulation. Thus the effect of spontaneous release on the appearance of EAD can be found if we compare the shapes of AP's produced after the last stimulation (shown in Figure 2A and 2B by arrows for $K_{m,ns(Ca)}$= 0.9 μM). When last stimulus is coincident in time with spontaneous release, the last AP has clearly expressed EAD. Otherwise, an increase in APD is observed.

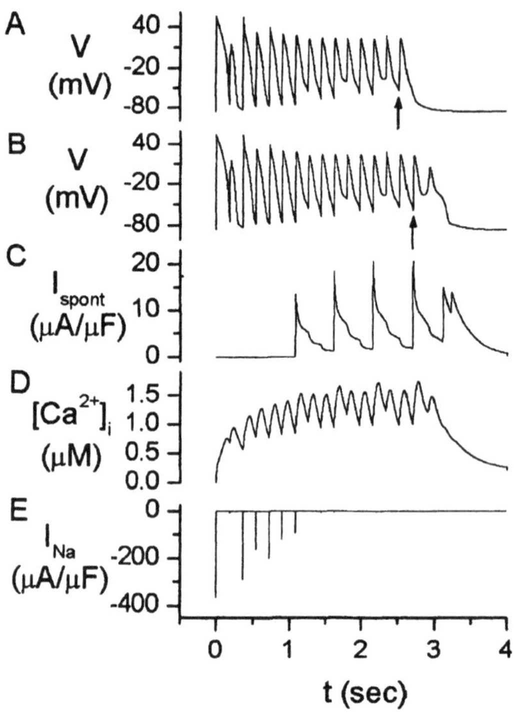

Figure 2. The cardiac cell model under high pacing rate (BCL=180 ms) and $K_{mns(Ca)}$.= 0.9 μM . The arrows in Figs. 2A and 2B show the moment of last stimulus application with respect to spontaneous release current peaks.

A. AP after last stimulation did not coincide in time with spontaneous release
B. AP after last stimulation at the time of spontaneous release
C. Spontaneous release current pattern in time
D. Intracellular free calcium concentration in time.
E. Changes of sodium current in time

The effect of the values of parameter $K_{mns(Ca)}$ on the appearance of EAD is illustrated in Figure 3.

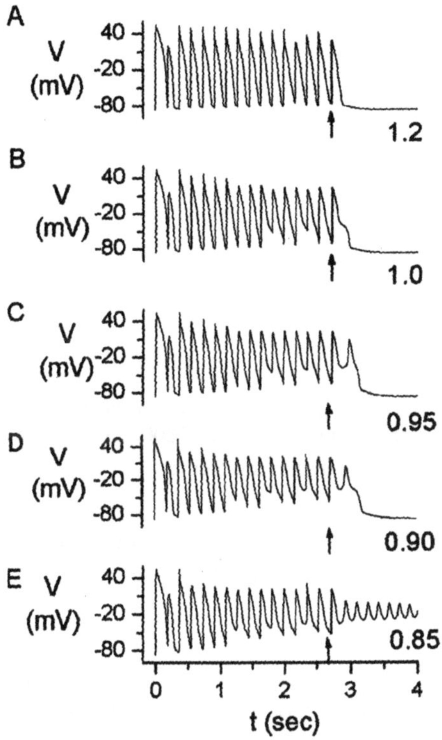

Figure 3. The formation of AP after the last stimulation for different values of $K_{m,ns(Ca)}$ installed at the beginning of each run of simulation In each case (A, B, C, and D) the last stimulus (denoted by arrows) coincides in time with spontaneous release current. BCL=180ms.

Here it is shown that the AP produced after the last stimulation (applied at the time of spontaneous release) has EAD only for $K_{mns(Ca)}$ equal 0.95 and 0.9 μM and adjusted from the very beginning of simulation. For higher values of that coefficient, some increase in the duration of the last AP is observed. For smaller values (Figure 3D), the APD is increased tremendously on account of multiple depolarizations, which occur during the repolarization phase. This situation may be called as multiple EAD and express a special kind of cell instability

Let us consider the last AP with EAD (see Figure 3C) shown in Figure 4 in large time scale designated as curve 0.9.

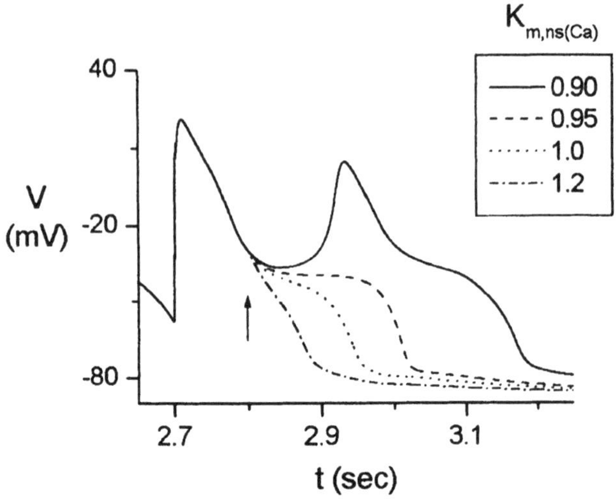

Figure 4. The effect of the $K_{m,ns(Ca)}$ perturbation on the shape of the AP repolarization phase. The arrow here indicates the moment of time (2800ms) at which $K_{m,ns(CA)}$ is changed from 0.9 μM to another value, as indicated in this figure.

We found here that between the two depolarized phases of AP there is a point for which dV/dt = 0. In this point the sum of inward and outward currents equals zero.

It is enough to introduce some small perturbation in the system to achieve the qualitative changes in voltage patterns. For example, changing the value of the $K_{mns(Ca)}$ from 0.9 to 1.0 μM approximately 40 ms before the time when equilibrium of inward and outward currents occurred converts the EAD into the prolonged AP (see Figure 4 curve 1.0). In both cases the APD is increased 2 to 3 times in comparison to that under rapid stimulation.

3.3 The excitation wave propagation in a 2D tissue model with decreased $K_{mns(Ca))}$.

The 2D tissue mathematical model is chosen as in [14] but the coefficient $K_{mns(Ca)}$ is changed from 1.2 or 1.0 μM to 0.9 μM in all cell models. The initial conditions for Ca^{2+} dynamics are: $[Ca^{2+}(0)]_{i,\,tot}$= 16.12 mM, $[Ca^{2+}(0)]_{NSR}$ = 1.3 mM, $[Ca^{2+}(0)]_{JSR}$= 6.78 mM.

The spiral wave is started by application of a premature beat (in the form of a vertical strip) close to the tail of the propagated rectilinear wave. The premature beat is applied at time t=258 ms from the initiation of a rectilinear wave front. The formation of the spiral wave is shown in Figure5a, picture A.

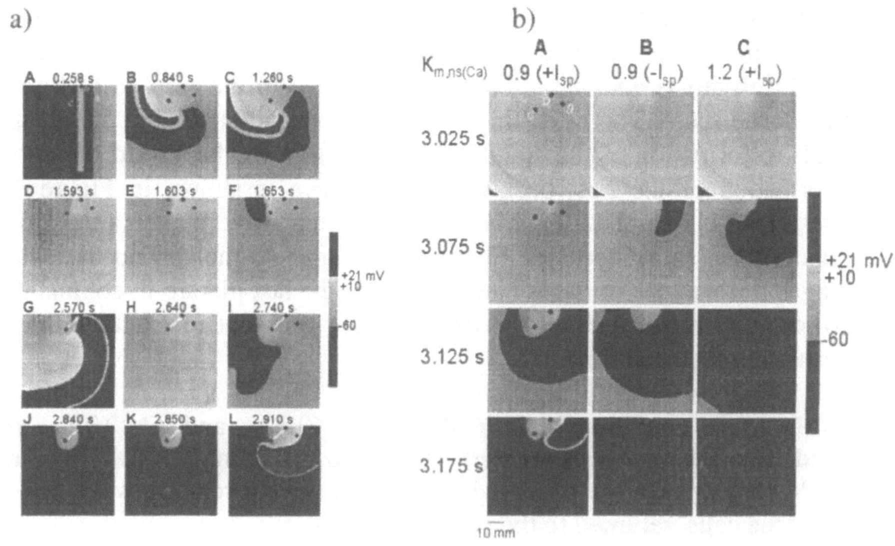

Figure 5. a) Ca^{2+} dynamics cause the conversion of stationary propagation to nonstationary, multiple stops of propagation with following formation of a nonstationary spiral wave rotating in the original and backward directions (A-K).
b) The elimination of backward rotating spiral waves.
In column A Figure 5 b, the continuation of the previous (Fig 5 a) spiral wave propagation is shown. Columns B and C illustrate the effect of blocking spontaneous release and changing the coefficient $K_{m,ns(Ca)}$ on elimination of the backward rotating spiral wave. All changes were made at time t=3000ms

After some transient period, which continues until t=840ms, the stationary spiral wave is established with the period of rotations T ~180 ms. (see Figure 5a, picture B). After ~420ms of stationary spiral wave rotation, the wave front twisted (become scalloping) and we observed the transfer to non-stationary spiral wave (Figure 5a, picture C). The character of wave propagation from the beginning to this time is similar to that obtained in [14] for $K_{mns(Ca)} = 1.2\ \mu M$.

The major distinctions are developed at time t=1595ms when all tissue cells are found at different levels of repolarization and the wave front stops propagating (Figure 5a, picture D). This situation recalls the case when the size of a tissue is not sufficient to provide spiral wave propagation with a given wavelength. Usually, in this case after some time all the tissue cells return to the resting potential and propagation is impossible without repeated stimulation. In our case we observed that not all the tissue cells go out of excitation and that after some time the new non-stationary spiral wave is formed rotating in the previous direction (Figure 5a, pictures E and F). After two turns of this spiral wave propagation, the described situation with the stop of propagation and initiation of a new wave in the same direction is repeated once more. The next (third) stop of propagation (Figure 5a, picture H) leads to formation of the spiral wave rotating in the backward direction. The formation of the first backward spiral wave is shown in Figure 5a, pictures H, I, J, K and L. This backward rotating wave makes almost one turn and after a short stop (see Figure 5b, column A) the second backward rotating spiral wave is initiated (see Figure 5b, the last picture in column A). After not quite a full turn all the tissue finally goes out of excitation non-uniformly in space and time.

When the initial conditions: $[Ca^{2+}(0)]_{I\ tot}$, $[Ca^{2+}(0)]_{NSR}$, and $[Ca^{2+}(0)]_{JSR}$ are changed from the above mentioned values to 20.85mM, 1.12 mM , and 6.94 mM respectively, we observed only one backward rotating wave before all the tissue cells returned to the resting potential.

The backward rotating spiral wave can be eliminated by blocking the spontaneous release of Ca^{2+} (see Figure 5, b, column B) or by changing the coefficient $K_{m,ns(Ca)}$ to the original value 1.2 μM (see Fig 5b, column C). All these changes were made at time 3000ms of the simulation run shown in Figure 5 (~ 125 ms before the initiation of the last backward rotating spiral wave). After some time the tissue returns to the resting potential (see Figure 5, b, the last two pictures in columns B and C).

4. Discussion

4.1 EAD and intracellular Ca^{2+} dynamics

The presented results showed that EAD is observed in the single cell model [14] after stopping of high rate stimulation at the moment when Ca^{2+} spontaneous release occurs. Figure 6 shows that with Luo-Rudy model [11] the same phenomenon can be observed.

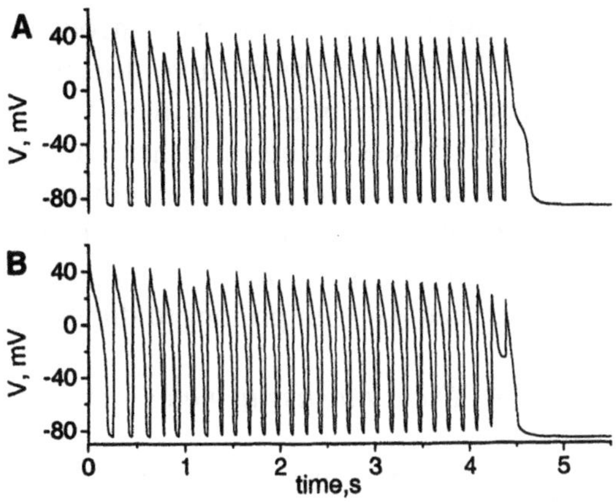

Figure 6. AP's pattern for cardiac cell model [11] under high pacing rate when the last stimulus coincides with SR spontaneous release.
A. For $K_{mns(CA)} = 1.2\ \mu M$ (normal), the increase of AP duration is observed.
B. For $K_{mns(CA)} = 0.9\ \mu M$, the single EAD appears.

The only difference is that in model [11] the $[Ca^{2+}]$ accumulation and SR spontaneous release required more time to occur than in model [14] due to different formulation of the calcium release (CICR) properties. Spontaneous Ca^{2+} release can cause the EAD only if it facilitates the appropriate increase of the $[Ca^{2+}]_i$ activation inward currents (e.g. nonspecific inward current $I_{ns(Ca)}$). The effect of SR spontaneous Ca^{2+} release on $I_{ns(Ca)}$ current in models [14] and [11] is amplified by decreasing the coefficient $K_{m,ns(Ca)}$ according the same expression in both models:

$$I_{ns(Ca)} = \frac{1}{\left(\frac{K_{m,ns(Ca)}}{[Ca^{2+}]_i}\right)^3 + 1}\left(\overline{I_{nsNa}} + \overline{I_{nsK}}\right)$$

The computer simulations results show (see Figure 3 and Figure 6) that decrease of this coefficient in the interval $1.2 - 0.95\ \mu M$ at the beginning of stimulation gives a smooth increase of the last AP duration. The further decrease leads to the transition to single (beginning with $K_{m,ns(Ca)} = 0.95\ \mu M$) and multiple EADs (beginning with $K_{m,ns(Ca)} = 0.85\ \mu M$).

According to equation (1), for the single cell the shape of the repolarization phase of the AP is determined by the time variations of the balance of inward and outward currents. Usually, during this phase of the AP, the outward currents prevailed over the inward ones. The exceptions are the cells where, after a notch, one can observe a small increase of membrane potential before a plateau phase. The appearance of the inward current $I_{ns(Ca)}$ as a result of Ca^{2+} overload in JSR and the successive spontaneous release facilitates the change of the current balance in the positive direction. Depending on the value of this misbalance and the time of its duration it is possible to obtain the increased APD or EAD (see Figure3). When outward currents are equal inward, theoretically it is possible to get an infinitely long AP, but this equilibrium state is usually non-stable. Indeed, it is enough to introduce a small distortion to the current balance in the negative direction (see Figure 4) as the EAD is changed to an AP with prolonged duration.

4.2 Propagation in 2D cardiac tissue model with decreased $K_{m,ns(Ca)}$

The origination of wave propagation in tissue models is obtained by application of external stimuli to some group of cells during a comparatively short time interval. If it is done before the tail of previous wave we initiate a spiral wave. Except for the very beginning, the wave propagates due to AP diffusion without external stimuli. After a comparatively short transient period, the stationary spiral wave is established with rotation period (~150-180 ms) significantly shorter than normal BCL (1000ms) and therefore the diffusion currents activate each cell in the area of stationary spiral wave propagation with high rate of repetition. That leads to Ca^{2+}accumulation in JSR of corresponding cells as was shown in [14] and in this paper (see Figure 2D) for a single cell under high pacing rate.

The stationary spiral wave propagation continues until the accumulation of Ca^{2+} would cause the overload and spontaneous release from JSR. From that moment on, due to the increased $I_{ns(Ca)}$ and I_{NaCa} currents, the APD of the cells with spontaneous release will be increased. Correspondingly the wavelength will non-uniformly change in space. Thus, the stationary spiral wave propagation becomes non-stationary with the characteristic scalloping of the wave front (see Figure 5a, picture C).

The wavelength increase may reach such a value that further wave propagation, at least temporarily, becomes impossible in the given tissue size (see Figure 5a, picture D). The wave stops and the conditions for appearance the APs with prolonged duration and EADs are created for some numbers of tissue cells. In Figure 7 A, B, C we illustrate this statement showing AP patterns obtained in course of simulation runs. These patterns were taken from different spatial locations denoted in Figure 5 by the points a, b, and c. The difference between the patterns begins to be visible with the first spontaneous release (t>1000ms). While in the pattern measured in point "a" there are no APs with EAD but only with prolonged APD, in patterns belonging to the points b and c, we observe EAD and more complex shapes of AP. The latter may represent the effect of the total diffusion current, on the cell in tissue. The diffusion currents, which may have inward or outward direction, are presented in cells, even after the stop of propagation and its values are comparable with other major currents.

The cells with EAD and prolonged APD formed some separate area of excited tissue surrounded by an area of resting cells going out of excitation see Fig 5a, pictures I, J, and K. When the cells, which went out of excitation, appear close to the border of this excited area supported by EAD, the new wave initiates in the direction of previous rotation (Figure 5a, picture F). or in the opposite (Figure 5a, pictures K and L). In both cases, the new spiral wave initiated in the area close to the tissue border and the border of unexcited tissue (see Figure 5a, picture K). The cells located in this region require minimal diffusion current for excitation because tissue borders, by definition, do not consume current.

It is necessary to note that inside this separate area we observe slow changes in repolarization levels distribution in space and time, which resemble wave propagation. In Figure 7D we show the AP patterns in the time interval from 2500 ms to 2900 ms in 14 different points located along the straight line connecting points "b" and "c" shown in Figure 5a.

The chosen time interval covered the time of the origination of the first backward rotating spiral wave. The analysis of these patterns allows us to conclude that repolarization levels change in some points of the straight line synchronously while in others some propagation with variable velocity ($\theta_{\max} = 15\ cm\ /\sec\ ; \theta_{over} = 12 cm/\sec$) is observed. The first two upper points demonstrate the AP with prolonged duration of AP. The initiation of these APs is caused by the wave front of previous spiral wave (Figure 5a, picture G). The next excitation of these points occurs due to the neighboring cells which are under a high level of repolarization (see Figure 5a, picture K and Figure 7B). In the rest of the points different forms of EAD are observed.

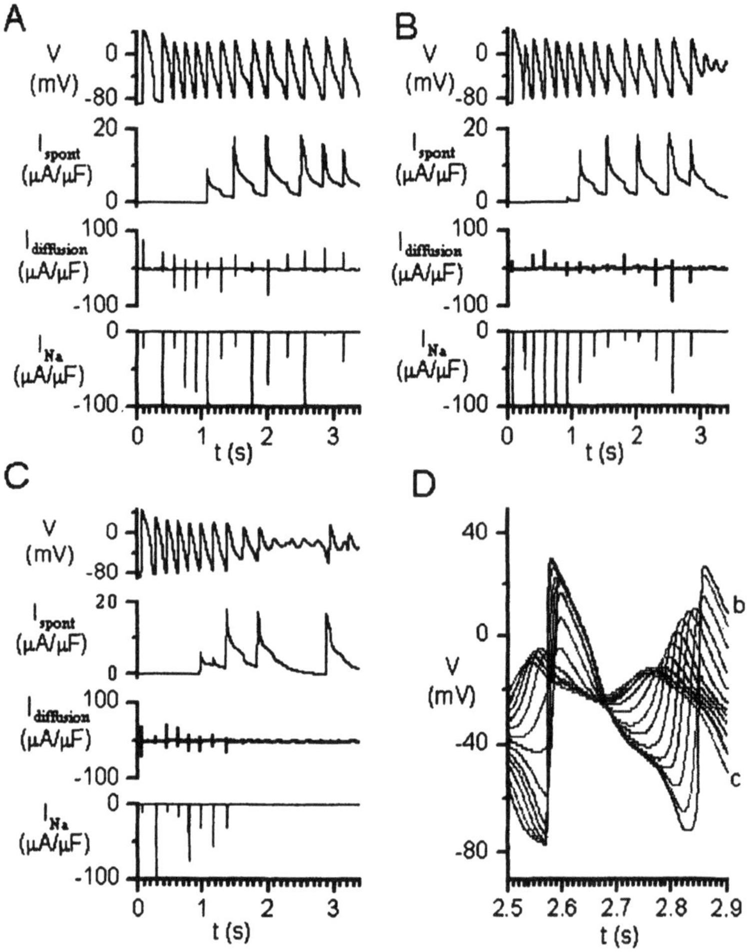

Figure 7 A, B, and C The APs and major currents patterns in time for the points "a", "b", and "c" (see Figure 5a). The locations of these points in space are chosen to cover different parts of the area with retarding repolarization.
D The APs are patterns taken in the time interval 2500-2900 msec of the computer simulation run. These patterns belong to 14 points located on a straight line connecting points "b" and "c".

4.3 *Limitation and implications of the study*

The limitations of the presented results reflect the limitations and assumptions used in the formulation of the modification of the Luo and Rudy model, which we used in our investigations without any significant changes. The detailed discussion of these limitations is presented in [14].

Because these limitations are connected with the study of the wave propagation in homogeneous isotropic 2D tissue, and with the assumptions that [Na] and [K] are constant in myoplasm, our results do not pretend to have a quantitative character. As soon as the model reproduces the major properties of Ca^{2+} dynamics: Ca^{2+} accumulation and spontaneous release from SR, we can consider that obtained results are qualitatively correct.

In spite of the fact that Ca^{2+} dynamics affect the membrane potential through various transmembrane currents, we restricted our study to investigation of the effect of Ca^{2+} dynamics on AP through the $I_{ns(Ca)}$ current. As it was shown in [19] the $I_{ns(Ca)}$ has the higher sensitivity to $[Ca^{2+}]_i$ among the other $[Ca^{2+}]_i$ dependent currents.

5. Conclusion

Our results show that:

1). The increase of $I_{ns(Ca}$ sensitivity$_)$ to $[Ca]_i$ under conditions of high pacing rate leads to prolongation of the AP duration in single cells due to Ca^{2+} accumulation and spontaneous release,

2). After stopping high rate pacing of a single cell, the single or multiple EAD emerge in the repolarization phase of the last AP, further increasing the duration of that AP,

3). In relation to wave propagation the increase of APD due to the Ca^{2+} spontaneous release cause the increase of a length of propagated spiral wave to such an extent that the size of the tissue becomes insufficient for further propagation.

4). The temporary block of propagation causes the appearance of cells with single and multiple EAD, which formed in the region of space with retarding repolarization. This area serves as the secondary source of consequent initiation of direct and backward spiral waves, when the surrounding tissue went out of excitation.

The new-presented findings not only expand our ideas about the possible regime of non-stationary wave propagation in the cardiac tissue model with intracellular Ca^{2+} dynamics, but also may initiate new physiological experiments to discover this phenomenon in real cardiac tissue.

Acknowledgment

The study was supported by National Institute of Health Grant SCOR in Sudden Cardiac Death P50HL52319. The supercomputer used for this investigation is provided by funding from National Energy Research

Scientific Computing Center, which is supported by the Office of Energy Research of the US Department of Energy under the contract No. DE-AC03-76SF00098. We also thank Eugene Chudin, Alan Garfinkel, and Zhilin Qu for useful discussions and comments.

References

[1] D. Noble. "A Modification of the Hodgkin-Huxley Equations Applicable to Purkinje Fiber Action and Pace-Maker Potentials". *J. Physiol.*, 160: 317-352, 1962.

[2] G. W. Beeler and H. Reuter. "Reconstruction of the Action Potential of Ventricular Myocardial Fibres". *J Physiol (Lond).*, 268(1): 177-210, 1997.

[3] C. Luo and Y. Rudy. "A Model of the Ventricular Cardiac Action Potential". *Circ. Res.*, 68: 1501-1526, 1991.

[4] R. Fitzhugh. "Mathematical Models of Excitation and Propagation in Nerve". In *Biological Engineering*, ed. H.P. Schwan, Chapter 1. McGraw-Hill, NY., 1969.

[5] F.J.L. Van Capelle, and D. Durrer. "Computer Simulation of Arrhythmias in a Network of Coupled Excitable Elements". *Circ. Res.*, 47(3): 454-466, 1980.

[6] A. Karma. "Spiral Breakup in Model Equations of Action Potential Propagation in Cardiac Tissue". *Phys. Rev. Lett.*, 71(7): 1103-1106, 1993.

[7] B.Y. Kogan, W.J. Karplus and M.G. Karpoukhin. "The Third-Order Action Potential Model for Computer Simulation of Electrical Wave Propagation in Cardiac Tissue". *Computer Simulation in Medicine.* ed. Power, H. and Hort R.T. Section 6. Computational Mechanics Publications, England, 1995.

[8] D. DiFrancesco. and D. Noble. "A Model of Cardiac Electrical Activity Incorporating Ionic Pumps and Concentrations Changes". *Phil. Trans. R. Soc.* London , B 307: 353-398, 1985.

[9] D.W. Hilgemann. D. Noble. "Excitation-contraction Coupling and Extracellular Calcium Transients in Rabbit Atrium: Reconstruction of Basic Cellular Mechanisms". *Proc. R. Soc. London*, B 230: 163-205, 1987.

[10] C. Nordin. "Computer Model of Membrane Currents and Intracellular Ca^{2+} Flax in Theisolated Guinea Pig Ventricular Myocyte". *Am.J. Physiol.*, 265: H2117-H2136, 1993.

[11] C. Luo and Y. Rudy. "A Dynamic Model of the Cardiac Ventricular Action Potential". *Circ. Res.*, 74: 1071-1096, 1994.

[12] J. Zeng, K.R. Laurita, D.S. Rosenbaum, Y. Rudy. "Two Components of the Delayed Rectifier K+ Current in Ventricular Myocytes of the Guinea Pig Type". *Circ. Res.*, 77: 140-152, 1995.

[13] M. Jafri, J. Rice, and R. Winslow. "Cardiac Ca^{2+} Dynamics: The Roles of Ryanodine Receptor Adaptation and Sarcoplasmic Reticulum Load". *Biophys. J.*, 74:1149 – 1168, 1998.

[14] E. Chudin, A. Garfinkel, J. Weiss, W. Karplus, B. Kogan. "Wave Propagation In Cardiac Tissue and Effects of Intracellular Calcium Dynamics". *Progress in Biophysics & Molecular Biology*, 69(2/3):225-2362, 1998.

[15] E. Chudin, J. Goldhaber, A. Garfinkel, J. Weiss, B. Kogan. "Intracellular Ca Dynamics and the Stability of Ventricular Tachycardia". *Biophis. J.*, 77:2930-2941, 1999.

[16] J. W. Bassani, W. Yuan and D. M. Bers. "Fractional SR Ca Release is Regulated by Trigger Ca and SR Ca Content in Cardiac Myocyte". *Am. J. Physiol.*, 268: C1313-C1319, 1995.

[17] M.D. Stern, M. C. Capogrossi, and E.G. Lakatta. "Spontaneous Calcium Release from the Sarcoplasmic Reticulum in Myocardial Cells: Mechanisms and Consequences". *Cell Calcium*, 9:247-256, 1988.

[18] G. Strang, "Numerical Analysis". *SIAM. J.*, 5: 526-517, 1969.

[19] B. Y. Kogan, W. J. Karplus, E. E. Chudin. "Heart Fibrillation and Parallel Supercomputers". *Proc. Int. Conference on Information and Control*, 3: 857-865, 1997.

[20] E. Chudin. "Instability of Cardiac Excitation Wave Propagation and Intracellular Calcium Dynamics". Dissertation for the PhD degree In Biomathematics, UCLA, 1999.

15

Simulation of Brain Hemodynamics: The Virtual Aneurysm

Daniel J. Valentino, Michael R. Harreld, Daren A. Lee and Gary R. Duckwiler

Please address correspondence to
Daniel J. Valentino
Associate Professor of Radiological Sciences,
David Geffen School of Medicine at UCLA
University of California, Los Angeles, CA 90095
dvalentino@mednet.ucla.edu

Summary: One of the most challenging applications of simulations in medicine is the use of Computational Fluid Dynamics (CFD) to simulate vascular hemodynamics and visualize the resulting complex flow patterns so that physicians can better understand and plan the treatment of patients with life-threatening vascular diseases such as brain aneurysms. It is now possible to acquire high-resolution images of the brain, and to subsequently create detailed 3D models of aneurysm geometry. CFD methods are used to calculate vascular flow patterns in the aneurysm, including fluid velocity and wall pressure and strain. Virtual-reality visualization techniques are then used to enable the physician to interactively explore vascular geometry and flow patterns in an immersive environment. The resulting simulation data and visualization techniques are helping medical researchers to develop new theories and tools to better organize and understand the complex anatomy and physiology of human brain vasculature.

1. Introduction

1.1 Dynamic simulation and virtual environments

Over the last decade, tremendous progress has been made in the use of modeling techniques and computer graphics to develop virtual environments that enable the interactive, immersive visualization of complex data sets [Bud1994, Bry1999]. More recently, interactive virtual environments have been combined with dynamic simulations (simulations based upon solutions of differential equations) to enable the virtual exploration of physical phenomena [Bry1999]. Such environments have been used, for example, in aerospace applications to visualize complex, three-dimensional, unsteady and non-linear fluid flow fields around airfoils and rotors [Modi2002]. In the medical field, virtual environments and virtual reality (VR) techniques have been used to train physicians and other healthcare professionals [Sat1996, Lee1999], as well as to improve surgical planning [Sat1996, Kik1996, Nak1997] and guide surgical navigation [Mas1998, Man1999, Jol2001].

In the mid 1990s, Walter J. Karplus and the authors became interested in the use of dynamic simulations embedded in a virtual environment to enhance the understanding and investigation of biomedical systems [Kar1996, Har1996]. In this chapter, we describe some of the approaches that have been developed to model the cerebrovasculature (the blood vessels in the brain), to simulate brain hemodynamics, and to visually interrogate the simulation results. The development of computational approaches to these issues involves a number of interesting problems in engineering and computer science. Furthermore, the computational approaches developed for cerebrovascular hemodynamics are applicable to a number of other areas, such as cardiology, and may be of use in other medical simulation problems.

1.2 Cerebrovascular Disease

1.2.1 Stroke

Stroke (hemorrhagic or ischemic) is the third leading cause of death in the United States, resulting in 160,000 deaths in 1998, and the leading cause of permanent disability [ASA2002]. Intracranial aneurysms, representing 25% of all hemorrhage-related strokes, are of specific interest to us because they are particularly difficult to treat and are associated with extremely high morbidity and mortality rates [Joh2001]. Each year, approximately 30,000 people in the US suffer from subarachnoid hemorrhages due to aneurysms. Almost 15% of these patients will die before reaching the hospital, over half will die within the first thirty days after hemorrhage, and half of the survivors will suffer a permanent disability that requires long-term nursing care [Ves1999, Joh2001, ASA2002]. Vascular hemodynamics are important in evaluating the extent and severity of disease, in planning treatment, monitoring progress, and evaluating outcomes; yet relatively little is known about the complex anatomy and hemodynamics of the cerebrovasculature [Tog1983, Ste1988, Had1994].

1.2.2 Brain aneurysms

An intracranial aneurysm is a sacculation (or outpouching) resulting from a weakening in the wall of a cerebral artery. Figure 1 illustrates the vascular anatomy of an intracranial aneurysm. The tube-like structure represents the artery, the arrow indicates the direction of blood flowing into the artery, and the spherical, balloon-like structure represents a saccular aneurysm on the side of the arterial wall. The larger part of the sac is known as the body of the aneurysm, and the part that attaches to the artery is known as the neck of the aneurysm. The top of the aneurysm is known as the dome, and is the part that is commonly believed to rupture as shown in Figure 1(b).

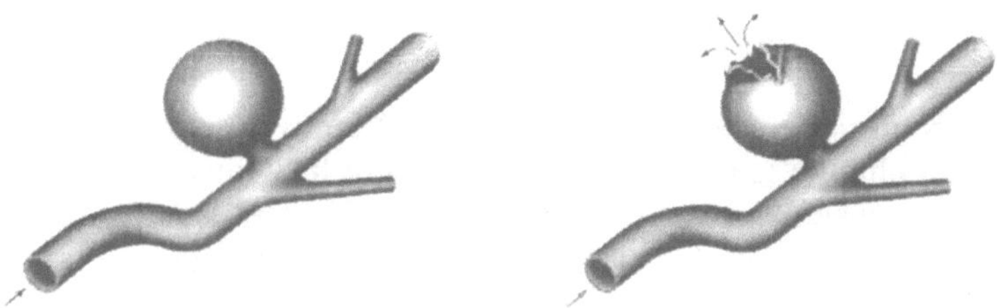

Figure 1. **Anatomy and Rupture of a Brain Aneurysm.** Figure 1(a) illustrates an idealized saccular aneurysm forming on the side of the arterial wall. Figure 1(b) illustrates the rupture of the dome of the aneurysm.

A surgical dissection of a real aneurysm is shown in Figure 2. As shown in this photograph, saccular aneurysms most commonly occur at the bifurcation of a blood vessel. A combination of flow rate and shear stress are believed to cause enlargement and rupture of aneurysms. Because the cerebrovasculature has high flow rates and shear stress, saccular aneurysms are highly prone to rupture. Both the immediate effects of the pressure caused by the hemorrhage and the secondary effects of the blood on the cerebral capillaries prevent oxygen from flowing to the brain and result in a stroke [Ves1999].

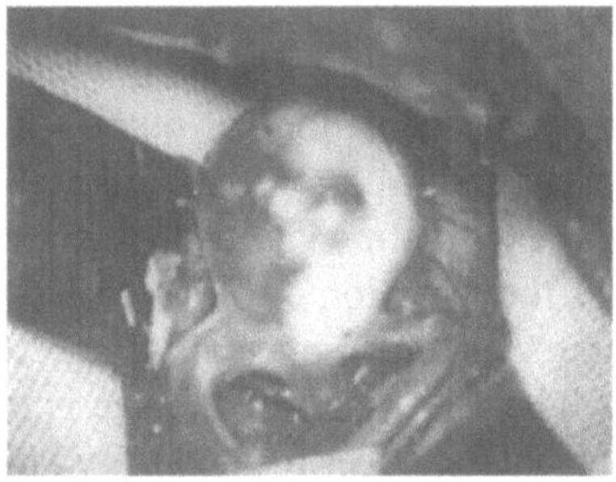

Figure 2. **Aneurysm at Surgery**. This is a photograph of a large aneurysm at the bifurcation of an artery. The aneurysm is the large sac on top of the parent artery (the larger vessel on the left) and the two branching vessels (in the center and on the right).

1.3 Treatment of Brain Aneurysms

After an aneurysm is detected, one option would be to observe it over time and wait until a change is observed before treating it. However, due to the risk and severe consequences of rupture, aneurysms are usually treated as soon as they are diagnosed. The most common treatment is surgical "clipping" in which a small clamp is placed around the neck of the aneurysm, thus preventing blood from flowing into the body (or lumen) of the aneurysm, and thereby preventing rupture. Unfortunately, some cerebral aneurysms are difficult to treat using standard surgical techniques. The surgical procedure itself requires manipulating critical brain tissue, thus risking damage to the brain. Furthermore, depending upon the location and geometry of the aneurysm, it may be difficult to place the clamp around the neck of the aneurysm. Lastly, due to the length and cost of surgical procedures and subsequent post-surgical care, the overall cost of surgical treatment is extremely high [Joh2001].

1.3.1 Endovascular Treatment

In the early 1990s, researchers at UCLA introduced a variety of techniques for endovascular occlusion of aneurysms, including balloons and other agents, and a detachable micro coil known as the Guglielmi Detachable Coil (GDC), [Gug1991]. As opposed to surgical treatment, GDC treatment involves "endovascular surgery" in which an interventional radiologist inserts a micro catheter into the patient's peripheral arterial system (such as the groin) and navigates it into the cerebrovasculature and then into the aneurysm as illustrated in Figure 3. The interventionalist then uses the micro catheter to introduce small platinum coils that fill the aneurysm and form a clot, thus preventing rupture of the aneurysm as shown in Figure 4.

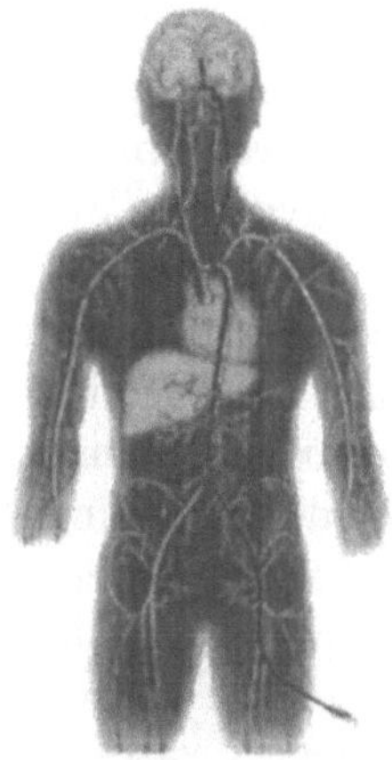

Figure 3. **Endovascular Navigation.** A catheter is inserted into a peripheral artery and navigated into the upper vasculature of the brain.

1.3.2 Limitations of Endovascular Treatment

The risk of endovascular surgery for treatment of intracranial aneurysms is substantially lower than the risk of surgery, but is nonetheless significant. Endovascular planning and treatment is complicated because physicians have limited knowledge of aneurysm anatomy and blood flow patterns. For example, typically more than one GDC is required to completely fill the aneurysm. However, it is often difficult for the interventionalist to predict the size and number of GDCs required to optimally treat the aneurysm because aneurysms vary greatly in their shapes, sizes and flow characteristics [Vil2000]

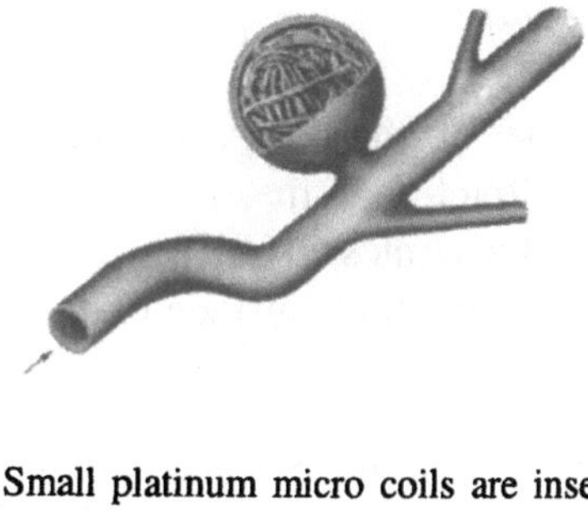

Figure 4. **GDC Treatment.** Small platinum micro coils are inserted into the body of the aneurysm to fill it and prevent rupture.

1.4 Challenges

Interactive, dynamic virtual environments are needed to improve pre-operative planning of interventional procedures and post-operative evaluation and follow-up. The key challenges are the development of image processing algorithms to extract geometric and physiologic information from the acquired data, the development and validation of simulation techniques to predict the hemodynamics of the cerebrovasculature, and the design and implementation of virtual reality (VR) environments to visualize and evaluate complex, three-dimensional anatomy and physiology. A related challenge is to create an effective educational environment for training interventionalists to better manage complex vascular diseases. Longer-term challenges include extending the simulation and training environment to simulate and evaluate patient outcomes and thereby improve clinical decision-making. The purpose of this chapter is to describe some of the techniques that have been developed to address the problem of simulating and visualizing complex vascular anatomy and hemodynamics. We will use a specific modeling, simulation and visualization environment known as the Virtual Aneurysm (VRA) to illustrate many of these techniques [Kar1996, Har1996].

2. Techniques

A number of approaches have been developed to enable physicians to visualize complex, three-dimensional anatomy and dynamically changing vascular hemodynamics. The primary components of such systems include the following: (1) image processing and analysis tools to acquire and model aneurysm geometry and other parameters from clinical data; (2) blood flow simulation tools to generate three-dimensional, dynamic simulations; (3) virtual reality and scientific visualization techniques to interactively display and interrogate the simulation results.

2.1 System Architecture

The specific components of the VRA include image acquisition, flow estimation, surface extraction, mesh generation, simulation using Computational Fluid Dynamics (CFD), and virtual reality based visualization. The flow estimation, surface extraction and mesh generation modules extract the necessary data from the patient's image data sets to model the aneurysm. The simulation module calculates the blood flow with the given aneurysm model parameters and stores the results in a solution database. A visualization module interacts with this solution database to request the data and uses a virtual environment to display the 3D, dynamic nature of the blood flow. Each of these modules is described in more detail in the following sections.

The VRA system was developed under the IRIX (Unix) Operating System (SGI, Mountain View, California) using the WorldToolKit VR Library (Sense8 Corporation, San Rafael, California) for management and rendering of the virtual environment. The image processing and numerical computations were performed on a 12-processor SUN 6800 symmetric multiprocessor (SUN Microsystems, Santa Clara, California). The simulation and database server was implemented on a 64-processor SGI Origin Graphics Supercomputer system, and the virtual world client was implemented on a Silicon Graphics Octane Workstation with a head-mounted display (Virtual Research, Inc) and a position tracking system (FASTRAK, Polhemus, Inc). A conventional 2D mouse was used to navigate through the aneurysm, manipulate virtual tools, and select the pull-down menus from the user interface.

2.2 *Data Acquisition*

The first step is to accurately diagnose the presence and evaluate the extent of a suspected aneurysm. Villablanca, et al, described a now common approach that uses X-ray Computed Tomographic Angiography (CTA) to acquire high-resolution, cross-sectional images of the cerebrovasculature. The CTA data is then rendered in 3D using interactive volume rendering. This approach has proven more effective than Digital Subtraction Angiography (DSA) for aneurysm detection and evaluation [Vil2000].

Once an aneurysm has been diagnosed, DSA is used to estimate bulk flow conditions and to later guide endovascular intervention. Other imaging modalities that are emerging as useful for characterizing aneurysms, and potentially for guiding interventions, include magnetic resonance angiography (MRA) and three-dimensional rotational angiography (3DXRA) [Sug2002].

2.3 *Flow Estimation*

Bulk flow measurements are required to compute the boundary conditions for the inflow vessels. These estimates of the bulk flow can be calculated from digital subtraction angiography (DSA) images acquired during the initial evaluation of the aneurysm or during endovascular treatment of the aneurysm. The instantaneous and mean bulk blood flow velocities at the inflow of the parent artery are computed from measurements of the changes in contrast density in the DSA images using video densitometric techniques.

The review article by Shpilfoygel, et al, contains an extensive survey and classification of the existing video densitometric techniques for estimating blood flow and velocity using DSA [Shp1999]. Although a variety of measurement techniques have been proposed, such as time of transit, conservation of mass, and distance-density methods, none of them work effectively on a wide range of images with differing flow velocities and image defect levels. Subsequent work by Shpilfoygel, et al, compared the most promising methods using a programmable flow phantom and clinically realistic flow and contrast injection conditions including imperfect mixing [Shp2000]. Based upon the results of these studies, the VRA uses a BMT-F method with extrapolation to obtain the lowest bias and the lowest variability both within and between studies in the estimation of bulk blood flow for the Virtual Aneurysm simulations.

2.4 *Surface Modeling*

2.4.1 *Surface Extraction Methods*

While volume rendering is useful in diagnosing the presence and extent of an aneurysm, it does not produce the geometric surface models that are required to simulate flow or to interact with the aneurysm in a virtual environment. The first step in surface modeling is the extraction of the geometric surface of the aneurysm. In the Virtual Aneurysm we use a combination of image processing and polygon generation techniques to extract the surfaces [Mar1999]. The VRA Surface Extraction module enables a physician to interactively view and manipulate the 3D data, and to select the aneurysm or region of interest (ROI) via a volume-bounding box. Next, region-based segmentation is used to isolate the aneurysm borders in the images, and the marching cubes algorithm is used to generate the surface mesh [Lor1987]. Mesh refinement techniques are then applied to reduce the number of redundant polygons (decimation) and reduce skew. For more fine-grained control, an expert user can optionally specify the parameters to the mesh generation process, such as the total number of triangles in the geometry or the maximum skew allowed.

2.4.2 *Tetrahedral Mesh Generation Methods*

After extracting the geometry of the aneurysm from the images, the next step is to create a grid of tetrahedrons (a tetrahedral mesh) that are used by the simulation engine to compute the fluid flow. The VRA uses a commercial tool (TGRID, Fluent, Inc) to create a mesh filled with tetrahedrons of various sizes and shapes, and to create the end caps for the inflow and outflow vessels. Both of these tasks can be fully automated using TGRID's scripting features. However, in some cases, errors in the surface mesh, such as overlapping polygons, can cause the tetrahedral creation process to fail or to produce an incorrect mesh. In these cases manual editing of the surface mesh is required until the generated tetrahedral mesh is acceptable. Figure 5 illustrates a typical aneurysm geometry model that was rendered using conventional Phong shading of the surface polygons.

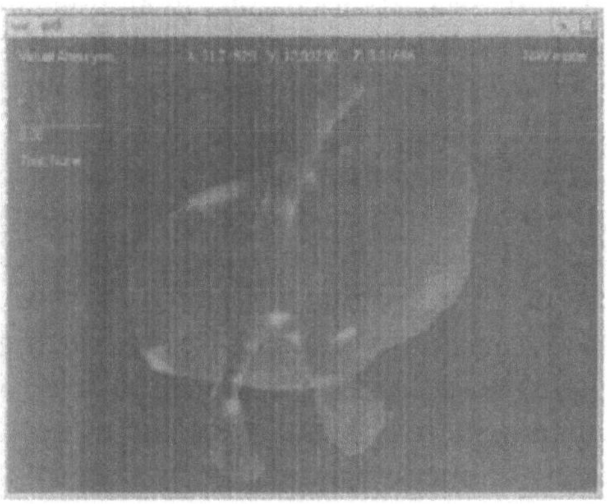

Figure 5. **Geometric model of the aneurysm.** The surface polygons that describe the geometric shape of the aneurysm were rendered using Phong shading.

2.5 Simulation

2.5.1 Simulation models

Despite the difficulty of accurately modeling and simulating flow in complex geometric shapes, several researchers have created computer-based models and simulations of the hemodynamics of intracranial aneurysms. Early efforts at modeling and simulation used two-dimensional, finite element methods to study flow patterns and particle paths. Perktold, et al, modeled a saccular aneurysm at the branching point of an arterial bifurcation [Per1989]. They simulated pulsatile flow under Newtonian and non-Newtonian conditions, and observed disturbed flow patterns in the bifurcation and relatively low flow circulation in the aneurysm. Burleson, et al, investigated lateral, side-wall cerebral aneurysms (side-wall aneurysms are those originating in the side of an arterial wall as opposed to the branching point of a bifurcating artery). They simulated three idealized side-wall aneurysms, including half-spherical, spherical, and pear-shaped [Bur1995]. These simulations indicated that maximal shear stresses and maximum pressures were typically in the downstream side of the aneurysm neck, rather than at the dome of the aneurysm, suggesting that the growth and rupture of the aneurysm occurs at the neck. Their model also indicated that aneurysm hemodynamics are highly dependent on a number of factors related to flow and geometric parameters, suggesting that the tendency for growth or thrombosis may be influenced by variations in the size or shape of the aneurysm.

More recent research has focused on modeling and simulating the hemodynamics of endovascular coiling procedures [Gro2001]. Different numbers of coils in a cerebral aneurysm were simulated to evaluate the effect of the percentage of aneurysm filling on pulsatile flow and pressure in the parent artery and the aneurysm lumen. Lastly, some researchers have focused on directly visualizing and quantifying complex hemodynamics using velocity-encoded magnetic resonance imaging [Buo1998]. Due to limitations of current MR imaging technology, these studies have been restricted to blood flow patterns in major vessels such as the coronary and carotid arteries. However, these approaches may eventually provide techniques to validate numerical simulations and to evaluate hemodynamics during or immediately after endovascular intervention.

Before leaving this section, we will mention that there are also mathematical models describing the growth and rupture of aneurysms. Early models used Laplace's law to define a linear relationship between the radius and wall tension. However, histopathological studies indicate that the aneurysm wall consists primarily of viscoelastic collagen, and is therefore subject to non-linear increases in tension as the radius increases. Hademenos, et al, developed a non-linear form of Laplace's law to describe the development and rupture of fusiform and saccular aneurysms [Had1994a, Had1994b]. This model predicts that the critical radius (the radius at which the aneurysm is predicted to rupture) is approximately 20% larger than predicted by the linear models, which better matches experimental observations.

2.5.2 Numerical Simulation Methods

All of the methods of simulating flow dynamics are based upon simplifications or assumptions for numerically solving the governing Navier-Stokes equation shown in Equation 1.

$$\rho\frac{\partial\vec{u}}{\partial t}+\rho(\vec{u}\bullet\vec{\nabla})\vec{u}=\rho\vec{f}-\vec{\nabla}p+\mu\nabla^{2}u$$

where

$\vec{u} \Rightarrow$ the velocity vector

$p \Rightarrow$ pressure

$t \Rightarrow$ time

$\vec{f} \Rightarrow$ the body force vector

$\rho \Rightarrow$ the density

$\mu \Rightarrow$ the viscosity

Equation 1: Navier-Stokes equation. This is the governing equation derived from conservation of momentum that describes the rate of change of velocity for an incompressible fluid. It is solved numerically using Finite Element Methods (FEM).

The VRA model makes a number of simplifying assumptions in order to obtain a numerical solution of the Navier-Stokes equation. (1) Flow is calculated for a three-dimensional, laminar, incompressible Newtonian fluid. We also experimented with non-Newtonian fluids (where the viscosity is a nonlinear function of the strain rate), with a dramatic increase in computation time but with no improvement in the simulation results. (2) The blood is contained in rigid vessel walls, imposing no-slip boundary conditions on the flow at the vessel walls (i.e., the velocity is required to be zero at the vessel wall). (3) Zero-pressure conditions are assumed at the outflow vessel to ensure that the velocity profile is stable. (4) The velocity and pressure are initially zero.

The surface geometry, tetrahedral mesh, and inflow conditions created by the previous modules are loaded into a Computational Fluid Dynamics (CFD) program that uses iterative numerical estimation techniques to solve the differential equations describing the fluid flow. The numerical estimation requires dividing the geometric shape into a discrete, finite element grid of irregularly spaced points (the tetrahedral mesh). These points are placed more densely in regions where the numerical solution is expected to change rapidly, and less densely in regions where the numerical solution is expected to be relatively uniform. Because the general solution is not known before the simulation is run, the final locations of the points are determined iteratively during the simulation.

The CFD program is used to compute the blood velocity and pressure at all points in space over the time interval of one heartbeat. In addition, the shear forces and pressure on the aneurysm walls are computed. The user can adjust the simulation by changing parameters such as the length and granularity of the simulation time or the density and viscosity of the simulated fluid. The simulation results are then inserted into a simulation database that is accessed by the virtual reality system.

2.6 *Interactive Visualization*

2.6.1 *Virtual environments*

Virtual environments provide an immersive setting for visualizing and investigating a number of different parameters depending upon the goals of the user. For example, Masutani, et al, created a system that combines three-dimensional models of vascular geometry obtained from X-ray computed tomographic angiography (CTA) or from magnetic resonance angiography (MRA) with intraoperative digital subtraction angiography (DSA) images [Mas1998].

The Virtual Aneurysm environment enables a physician to interactively explore the dynamic behavior of the circulation patterns in the aneurysm as they vary in time [Har1996]. The variables that can be investigated include flow rates and patterns in the lumen of the artery and body of the aneurysm, and pressure and shear stress on the wall of the artery and aneurysm.

2.6.2 *Virtual Aneurysm Tools*

The VRA visualization module provides dynamic, virtual reality-based tools that enable an interventional radiologist to obtain information to help diagnose the presence and extent of an aneurysm, and to assess various methods of treatment. These tools can be classified into two main groups: (1) global flow circulation tools and (2) local primitive data tools.

The flow circulation tools are used to display the circulation patterns of time-dependent flow fields. As the name implies, the global flow circulation tools display a global representation of the blood flow. Example tools include streak-lines, particle paths, and ribbons. By using these types of tools, the physician can visualize where the blood flows and the path taken to get there. The local primitive data tools are used to examine the flow field at specific (localized) points in the aneurysm. Example tools include color-mapped cutting planes, vector planes, and contour skins showing pressure or shear stress. By using these types of tools, the physician can examine the local aneurysm hemodynamics in great detail.

Figure 6 contains one frame from an interactive, real-time virtual reality visualization that showed interesting flow characteristics at the intersection of an aneurysm and two inlet vessels. A cutting plane representing the flow velocity vector has been placed to intersect the inlet and outlet flow fields. We can see that the two inlet flows are causing two vortices of flow in opposite directions.

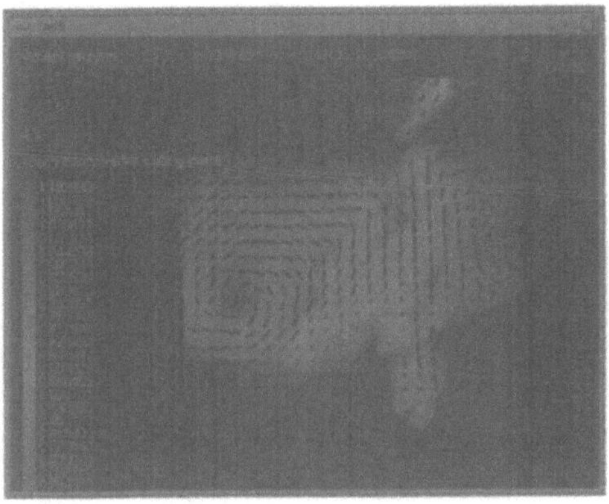

Figure 6. **Cutting Plane Tool.** The VRA enables the interventionalist to move a cutting plane showing the velocity vectors in the plane at different locations in the aneurysm. Note the circular flow patterns seen in the lumen of the aneurysm.

We have also developed techniques to visualize volumetric images and static or time-dependent vector fields over the Web using VRML and Java [Mar1999]. Physicians at different locations can simultaneously visualize a data set, and one of them can use virtual tools to illustrate observed hemodynamic patterns or indicate a proposed interventional procedure.

2.6.3 Feature Extraction

As we have seen, it is possible to create virtual environments that enable the interrogation of complex anatomic structures and hemodynamic patterns. Our early experience with such data and feature-rich virtual environments demonstrated that the physicians were often overwhelmed by the complexity of the data (and the corresponding interrogative tools), and therefore they spent most of their time searching for useful features in the data set [Har1996]. In order to help physicians identify important hemodynamic patterns, Lee, et al, developed automated algorithms to extract characteristic patterns in three-dimensional, dynamic flow fields [Lee2001]. The algorithm decomposes hemodynamic patterns into "flow skeletons" that can be visualized individually or superimposed to formulate an understanding of the flow patterns in the aneurysm. Figure 7 contains one frame from a visualization of a single flow skeleton in a simulation. The flow skeleton consists of a set of particle paths traced for 30 out of 80 time steps. The aneurysm boundary was rendered in wire frame, and the particles were seeded at the larger inlet vessel. We see the large circular flow with many particles ending near the aneurysm boundary, perhaps building up pressure/strain in that area. The squares represent the seed positions of the particles; the spheres the end positions.

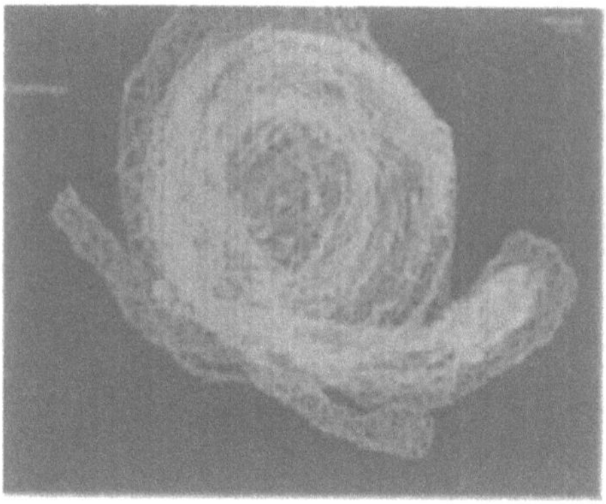

Figure 7. **Flow components.** This is a lateral view of the aneurysm and a flow skeleton showing a set of particle paths traced for 30 time steps. This flow skeleton shows those particles accumulating near the aneurysm boundary, indicating a region of potentially high strain and rupture.

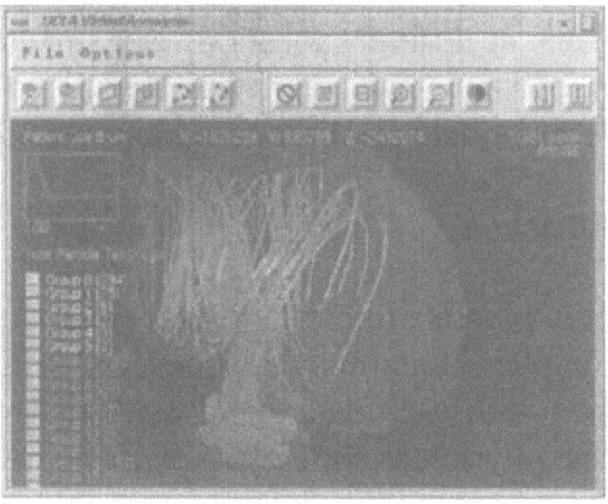

Figure 8. **Flow Components.** This is an anterior view of a flow skeleton showing a set of particle paths, again traced for 30 time steps. This view clearly shows those particles that accumulate on the left side of the aneurysm.

Figure 8 shows a similar characteristic flow pattern from another viewpoint, illustrating more clearly how these particles accumulate along the left wall of the aneurysm. Note that this is only one of several characteristic patterns seen in this simulation. For a complete description of the flow patterns that were identified see Lee, et al, [Lee2001].

The general approach taken was to seed the flow with massless particles, trace their paths for a given time, and then group the particle paths together into the flow skeleton. The skeleton was generated in three steps. First, for each set of seed points, the corresponding particle paths were grouped together based on their global behavior using Voronoi clustering and further refined based on local characteristics using region growing techniques. Next, the groups of particle paths were reduced to individual skeletons using smoothing and region growing techniques. Lastly, these individual skeletons were merged together to form the flow skeleton.

2.6.4 Simulating treatment and predicting treatment outcomes

It is also possible to simulate aneurysm treatment by surgical clipping and by endovascular coiling. For example, to simulate surgical clipping, a cutting plane can be intersected with the aneurysm surface to create a new geometric object with all or part of the body of the aneurysm removed. To simulate coil treatment, one or more spherical objects can be intersected with the original aneurysm surface to create a new object for which all or part of the body of the aneurysm is occluded. The simulation process is repeated for this new mesh and then the effects of the treatment can be visualized.

3. Conclusions

3.1 Summary

As the scientific literature and examples from the Virtual Aneurysm have illustrated, simulation and virtual reality-based dynamic visualization environments are useful in understanding the anatomy and patterns of flow of complex hemodynamic systems. Such environments have been used to identify potential points of rupture, to investigate the origin of these rupture zones, and to evaluate treatment options. In the future, such environments will enable the prediction of aneurysm growth and rupture, as well as the simulation of the most likely outcome of particular approaches. This will be an enormous aid in determining the appropriate therapeutic treatment. It is interesting to note that, as predicted, the shape of the aneurysm heavily influences its hemodynamics. However, the simulations produced flow patterns and resulting shear stresses that were sometimes unexpected, pointing to a need to reexamine some of our understanding of intracranial aneurysm rupture. We believe that environments such as the Virtual Aneurysm will eventually provide the functionality needed to enable the interactive investigation of these questions, however, a number of challenging computational issues remain to be addressed.

3.2 Remaining Challenges

The use of virtual environments may revolutionize clinical treatment planning and procedures, but such systems present significant computational challenges that cross academic disciplines ranging from mathematical modeling to image understanding to scientific visualization. We will briefly discuss some of those challenges here.

3.2.1 Simulation Validation

Perhaps one of greatest challenges is to validate and improve the accuracy of simulation of complex hemodynamic systems. The error rates of current simulations are difficult to assess due to the lack of accurate validation techniques. Recently, Tateshima, et al, developed a new in vitro method to analyze flow hemodynamics over one cardiac cycle [Tat2001]. Accurate velocity measurements were obtained in an acrylic aneurysm model using Laser Doppler Velocimetry. Such data can be compared to simulation results to identify the magnitude and source of simulation errors. Once the errors are better understood, simulation modeling can focus on resolving them using a variety of techniques. For example, since the geometric shape of the aneurysm plays a critical role in determining the hemodynamics, the mesh resolution can be increased in critical areas of the aneurysm. Numerical instability can be addressed using a combination of adaptive meshing and improved numerical estimation of the Navier-Stokes equations.

3.2.2 Virtual Environments

Another challenge is to improve the ability to navigate in a virtual environment. In addition to developing more useful virtual tools, an important problem is the automation of mundane visualization tasks. This automation should enable the physician to concentrate on aneurysm diagnosis and treatment assessment rather than the manipulation of the user interface controls. The recent work by Lee, et al, is promising because the approach is able to automatically identify and classify many patterns of flow in the simulation data, and display the resulting flow skeletons individually or collectively [Lee2001]. Ongoing work is needed to determine which of these patterns is important in determining treatment options.

3.2.3 Computational Complexity

Another significant challenge is the computational cost of hemodynamic simulations. Interactive simulation of the time-course of aneurysm hemodynamics requires intensive computation that will likely require massively parallel computing resources. Even more challenging is the significant biological, mathematics and computer science expertise required to improve and validate the accuracy of hemodynamic simulation. For example, an exciting but open question is how to combine mathematical models of aneurysm growth and rupture (e.g., Had1994a and Had1994b) with the Navier-Stokes models describing the flow patterns. The emerging discipline of Computational Neurobiology may bring about the integration of knowledge required, and thereby enable researchers to study extremely this and other challenging simulation problems in the medical domain.

In Memoriam

This work is dedicated to the memory of Professor Walter Karplus. Walter became interested in developing a virtual environment for the simulation and visualization of brain blood flow after seeing a presentation by Professor Fernando Vineula in the Department of Radiological Sciences. Fernando introduced Walter to two of the authors (DJV and GRD), which led to nearly a decade of collaboration on the Virtual Aneurysm. This collaboration produced some of the earliest scientific work in this area as evidenced by numerous presentations, manuscripts and grants. More importantly, this collaboration produced several masters and Ph.D. students, including two of the authors of this work (MRH and DAL). Most importantly, it led to a long relationship between Walter and the authors that we only wish could have lasted longer. He was our colleague, our mentor and our friend; we miss him greatly.

Acknowledgments

We would like to thank the faculty and staff of the Neuroendovascular Section and the Rigler Laboratory in the UCLA Department of Radiological Sciences for their help in the research and development of the Virtual Aneurysm over the past decade. The Virtual Aneurysm is supported in part by NIH/NCRR Resource Grant RR13642, and NIH Grant MH65166, as well as by industry grants from SUN Microsystems and Philips Medical Systems.

References

[1] American Stroke Association, Stroke Facts, 2002. http://www.strokeassociation.org/, [ASA2002]

[2] S. Bryson, D. Kenwright, M. Cox, D. Ellsworth, R. Haimes. "Visually Exploring Gigabyte Data Sets in Real Time". *Communications of the ACM*, 42(8):83-90, August, 1999. [Bry1999]

[3] M.H. Buonocore. "Visualizing blood flow patterns using streamlines, arrows, and particle paths". *Magn Reson Med*, 40(2):210-26, 1998. [Buo1995]

[4] A.C. Burleson, C.M. Strother, V.T. Turitto. "Computer modeling of intracranial saccular and lateral aneurysms for the study of their hemodynamics". *Neurosurgery*, 37(4):774-82, 1995. [Bur1995]

[5] C. Groden, J. Laudan, S. Gatchell, H. Zeumer. "Three-dimensional pulsatile flow simulation before and after endovascular coil embolization of a terminal cerebral aneurysm". *J Cereb Blood Flow Metab*, 21(12):1464-71, 2001. [Gro2001]

[6] G. Guglielmi, F. Vinuela, I. Sepetka, V. Macellari. "Electrothrombosis of Saccular Aneurysms via an Endovascular Approach". *Journal of Neurosurgery*, 75: 1-14, 1991. [Gug1991]

[7] G.J. Hademenos, T. Massoud, D.J. Valentino, G. Duckwiler, F. Vinuela. "A nonlinear mathematical model for the development and rupture of intracranial saccular aneurysms". *Neurol Res*, 16(5):376-84, 1994. [Had1994a]

[8] G.J. Hademenos, T. Massoud, D.J. Valentino, G. Duckwiler, F. Vinuela. "A nonlinear mathematical model for the development and rupture of intracranial fusiform aneurysms". *Neurol Res*, 16(6):433-8, 1994. [Had1994b]

[9] M.R. Harreld, D.J. Valentino, W.J. Karplus. "The virtual aneurysm. Virtual reality in endovascular therapy". *Stud Health Technol Inform*, 29:12-20, 1996. [Har1996]

[10] J. Helman, L. Hesselink. "Visualizing vector field topology in fluid flows". *IEEE Computer Graph. Applic.* 11(3):36-40, May, 1991. [Hel1991]

[11] S.C. Johnston, S. Zhao, R.A. Dudley, M.F. Berman, D.R. Gress. "Treatment of unruptured cerebral aneurysms in California". *Stroke*, 32(3):597-605, 2001. [Joh2001]

[12] F.A. Jolesz, A. Nabavi, R. Kikinis. "Integration of Interventional MRI with Computer Assisted Surgery". *JMRI*, 13(1):69-77, 2001. [Jol2001]

[13] W.J. Karplus, M.R. Harreld, D.J. Valentino. "Simulation and visualization of the fluid flow field in aneurysms: a virtual environments methodology". *MHS96: Proceedings of the Seventh International Symposium on Micro Machine and Human Science*, Nagoya, Japan, 35-41, Oct 2-4, 1996. [Kar1996]

[14] R. Kikinis, G. Langham, T.M. Moriarty, M.R. Moore, E. Alexander, M. Matsumae, W.E. Lorensen, P.M. Black, F.A. Jolesz. "Computer-assisted interactive three-dimensional planning for Neurosurgical procedures". *Neurosurgery*, 38(4):640-651, 1996. [Kik1996]

[15] H. Kiyosue, S. Tanoue, M. Okahara, Y. Hori, T. Nakamura, H. Nagatomi, H. Mori. "Anatomic features predictive of complete aneurysm occlusion can be determined with three-dimensional digital subtraction angiography". *Am J Neuroradiol*, 23(7):1206-13, 2002. [Kiy2002]

[16] D. Lee, M. Woo, D. Vredevoe, J. Kimmick, W.J. Karplus, D.J. Valentino. "Ophthalmoscopic examination training using virtual reality". *Virtual Reality*, 4(3):184-91, 1999. [Lee1999]

[17] D. Lee, D.J. Valentino, G.R. Duckwiler, W.J. Karplus. "Automatic skeleton generation for visualizing 3D, time-dependent fluid flows: application to the virtual aneurysm". *Stud Health Technol Inform*, 81:272-8, 2001. [Lee2001]

[18] W. Lorensen, H. Cline. "Marching Cubes: A high-resolution three-dimensional surface construction algorithm". *ACM Computer Graphics*, 21, 163-169, 1987. [Lor1987]

[19] M.E. Maniscalco-Theberge, D.C. Elliott. "Virtual reality, robotics,

and other wizardry in 21st century trauma care". *Surg Clin North Am*, 79(6):1241-8, 1999. [Man1999]

[20] B.S. Marovic, D.J. Valentino, W.J. Karplus, Z. Jovanovic. "Interactive visualization of 3D fields and images on inexpensive workstations using VRML". *Stud Health Technol Inform*, 62:219-20, 1999. [Mar1999]

[21] Y. Masutani, T. Dohi, F. Yamane, H. Iseki, K. Takakura. "Augmented reality visualization system for intravascular neurosurgery". *Comput Aided Surg*, 3(5):239-47, 1998. [Mas1998]

[22] A. Modi, L Long, N. Sezer-Uzol, P. Plassmann. "Scalable Computational Steering System for Vizualization of Large-Scale CFD Simulations". In the *Proceedings of the AIAA Fluids Conference*, St. Louis, 2002. [Modi2002]

[23] S. Nakajima, H. Atsumi, A.H. Bhaleroa, F.A. Jolesz, R. Kikinis, T. Yoshimine, T.M. Moriarty, P.E. Steig. "Computer-assisted Surgical Planning for Cerebrovascular Neurosurgery". *Neurosurgery*, 41(2):403-409, 1997. [Nak1997]

[24] K. Perktold, R. Peter, M. Resch. "Pulsatile non-Newtonian blood flow simulation through a bifurcation with an aneurysm". *Biorheology*, 26(6):1011-30,1989. [Per1989]

[25] R.M. Satava. "Three-Dimensional Imaging and Image Displays: Surgical Application of Advanced Technologies". *Semin Laparosc Surg*, 3(3):193-198, 1996. [Sat1996]

[26] S.D. Shpilfoygel, R. Jahan, R.A. Close, G.R. Duckwiler, D.J. Valentino. "Comparison of methods for instantaneous angiographic blood flow measurement". *Med Phys*, 26(6):862-71, 1999. [Shp1999]

[27] S.D. Shpilfoygel, R.A. Close RA, D.J. Valentino, G.R. Duckwiler. "X-ray videodensitometric methods for blood flow and velocity measurement: a critical review of literature". *Med Phys*, 27(9):2008-23, 2000. [Shp2000]

[28] H.J. Steiger, D.W. Liepsch, A. Poll, H.J. Reulen. "Hemodynamic stress in terminal saccular aneurysms: a Laser-Doppler study". *Heart Vessels*, 4:162-169, 1988. [Ste1998]

[29] T. Sugahara, Y. Korogi, K. Nakashima, S. Hamatake, S. Honda, M. Takahashi. "Comparison of 2D and 3D Digital Subtraction Angiography in the Evaluation of Intracranial Aneurysms". *Am J*

Neuroradiol, 23(9):1545-52, 2002. [Sug2002]

[30] D. Sujudi, R. Haimes "Identification of swirling flow in 3D vector fields". in the *Proceedings of the AIAA Computational Fluid Dynamics Conference*, San Diego, June 19-22, American Institute of Aeronautics and Astronautics, Washington, D.C., 792-799, 1995. [Suj1995]

[31] F. Tognetti, P. Limoni, C. Testa. "Aneurysm growth and hemodynamic stress". *Surg Neurol*, 20:74-78, 1983. [Tog1983]

[32] P.M. Vespa, Y.P. Gobin. "Endovascular treatment and neurointensive care of ruptured aneurysms". *Crit Care Clin*, 15(4):667-84, 1999. [Ves1999]

[33] J.P. Villablanca, N. Martin, R. Jahan, Y.P. Gobin, G.R. Duckwiler, J. Bentson ,M. Hardart, D. Coiteiro, J. Sayre, F. Vinuela. "Volume-rendered helical computerized tomograhy angiography in the detection and characterization of intracranial aneurysms". *J Neurosurg*, 93(2):254-64, 2000. [Vil2000]

16

Evolving Neural Networks: Selected Medical Applications and the Effects of Variation Operators

David B. Fogel
Natural Selection, Inc.
3333 N. Torrey Pines Ct., Suite 200, La Jolla, CA 92037
dfogel@natural-selection.com

Summary. Evolutionary algorithms can be used to train and design neural networks for medical applications. This paper reviews some recent efforts in breast cancer detection using evolutionary neural networks. The results obtained are discussed in relation to other methods for analyzing similar data. Additional basic research data are presented that investigate the use of alternative forms of variation on neural networks (e.g., mutation and recombination). Mention is given to the inspiration that Walter Karplus provided to the author in applying computational intelligence methods to practical problems in medicine and other disciplines.

1. Introduction

Neural networks and evolutionary algorithms provide computational tools for addressing pattern recognition problems. Walter Karplus was a strong proponent of these tools, particularly as applied to medical domains. This paper provides a brief review of the author's efforts in this direction as well as results from a study comparing alternative evolutionary variation operators in optimizing neural networks for function optimization. The fundamental technologies of neural and evolutionary computation are summarized and the experiments undertaken are detailed.

1.1 Artificial Neural Networks

Artificial neural networks (or simply *neural networks*) are computer algorithms based loosely on modeling the neuronal structure of natural organisms. They are stimulus-response transfer functions that accept some input and yield some output. They are used typically to learn an input-output mapping over a set of examples. For example, the input can be radiographic features from mammograms, with the output being a decision concerning the likelihood of malignancy.

Neural networks are parallel processing structures consisting of nonlinear processing elements interconnected by fixed or variable weights. They are quite versatile, for they can be constructed to generate arbitrarily complex decision regions for stimulus-response pairs. That is, in general, if given sufficient complexity, there exists a neural network that will map every input pattern to its appropriate output pattern, so long as the input-output mapping is not one-to-many (i.e., the same input having varying output). Neural networks are therefore well suited for use as detectors and classifiers. The classic pattern recognition algorithms require assumptions concerning the underlying statistics of the environment. Neural networks, in contrast, are nonparametric and can effectively address a broader class of problems [1].

Multilayer perceptrons, also sometimes described as *feed forward networks*, are probably the most common architecture used in supervised learning applications (where exemplar patterns are available for training). Each computational node sums N weighted inputs, subtracts a threshold value, and passes the result through a logistic (sigmoid) function. Single perceptrons (i.e., feed forward networks consisting of a single input node) form decision regions separated by a hyperplane. If the input from the given different data classes are linearly separable, a hyperplane can be positioned between the classes by adjusting the weights and bias terms. If the inputs are not linearly separable, containing overlapping distributions, a least mean square (LMS) solution is typically generated to minimize the mean squared error between the calculated output of the network and the actual desired output. While single perceptrons can generate hyperplane boundaries, perceptrons with a hidden layer of processing nodes have been proven to be capable of approximating any measurable function [2], indicating their broad utility for addressing general pattern recognition problems.

Another versatile neural network architecture is the *radial basis function network*. Rather than partitioning the available data using hyperplanes, the radial basis function network clusters available data, often with the use of approximate Gaussian density functions. The network comprises an input layer of nodes corresponding to the input feature dimension, a single hidden layer of nodes with computational properties described below, and output nodes that perform linear combinations on the hidden nodes. Each connection between an input node and hidden node carries two variable parameters corresponding to a mean and standard deviation. Poggio and

Girosi [3] proved that linear combinations of these near-Gaussian density functions can be constructed to approximate any measurable function. Therefore, like the multilayer perceptron, radial basis functions are universal function approximators.

Given a network architecture (i.e., type of network, the number of nodes in each layer, the connections between the nodes, and so forth), and a training set of input patterns, the collection of variable weights determines the output of the network to each presented pattern. The error between the actual output of the network and the desired target output defines a response surface over an n-dimensional hyperspace, where there are n parameters (e.g., weights) to be adapted. Multilayer feed forward perceptrons are the most commonly selected architecture and training these networks can be accomplished through a *back propagation* algorithm, which implements a gradient search over the error response surface for the set of weights that minimizes the sum of the squared error between the actual and target values.

Although the use of back propagation is common in neural network applications, it is quite limiting. This procedure is mathematically tractable and provides guaranteed convergence, but only to a locally optimal solution. Even if the network's topology provides sufficient complexity to completely solve the given pattern recognition task, the back propagation method may be incapable of discovering an appropriate set of weights to accomplish the task. When this occurs, the operator has several options: (1) accept suboptimal performance, (2) restart the procedure and try again, (3) use ad hoc tricks, such as adding noise to the exemplars, (4) collect new data and retrain, or (5) add degrees of freedom to the network by increasing the number of nodes and/or connections.

Only this last approach, adding more degrees of freedom to the network, is guaranteed to give adequate performance on the training set, provided sufficient nodes and layers are available. Yet this also presents problems to the designer of the network, for any function can map any measurable domain to its corresponding range if given sufficient degrees of freedom. Unfortunately, such overfit functions generally provide very poor performance during validation on independently acquired data. Such anomalies are encountered commonly in regression analysis, statistical model building, and system identification. Assessing the proper trade-off between the goodness-of-fit to the data and the required degrees of freedom requires information criteria (e.g., Akaike's information criterion, minimum description length principle, predicted squared error, or others). By relying on the back propagation method, the designer almost inevitably accepts that the resulting network will not satisfy the maxim of parsimony, simply because of the defective nature of the training procedure itself. The problems of local convergence with the back propagation algorithm indicate the desirability of training with stochastic optimization methods such as simulated evolution, which can provide convergence to globally optimal solutions.

2. Evolving Neural Networks

Natural evolution is a population-based optimization process. Simulating this process on a computer results in stochastic optimization algorithms that can often outperform classical methods of optimization when applied to difficult real-world problems. Historically, there have been three main avenues of research in simulated evolution: evolutionary programming, evolution strategies, and genetic algorithms. The methods are broadly similar in that each maintains a population of trial solutions, imposes random changes to those solutions, and incorporates the use of selection to determine which solutions to maintain into future generations and which to remove from the pool of trials. The methods differ in the types of random changes that are used and the methods for selecting successful trials. Fogel [4] provides a review of the similarities and differences between these procedures. The methods have been shown to possess asymptotic global convergence properties, and in some cases the techniques can be shown to have geometric rates of error convergence [4], making them attractive for function optimization problems.

The procedures generally proceed as follows. A problem to be solved is cast in the form of an objective function that describes the worth of alternative solutions. Without loss of generality, suppose that the task is to find the solution that minimizes the objective function. A collection (population) of trial solutions is selected at random from some feasible range across the available parameters. Each solution is scored with respect to the objective function. The solutions (parents) are then mutated and/or recombined with other solutions in order to create new trials (offspring). These offspring are also scored with respect to the objective function and a subset of the parents and offspring are selected to become parents of the next iteration (generation) based on their relative performance. Those with superior performance are given a greater chance of being selected than are those of inferior quality. Fogel [4] details examples of evolutionary algorithms applied to a wide range of problems, including designing neural networks, and these techniques have been applied to real problems in areas including pharmaceutical design, factory scheduling, freeway onramp metering, and various medical domains.

Optimizing neural networks through simulated evolution offers not only a superior search for appropriate network parameters, but the evolution can be used to adjust the network's topology simultaneously. By mutating both the structure of the network and its associated parameters (weights), a very fast search can be made for a truly robust design. This frees the operator from having to preselect a topology and then searching for the best weights under that constraint. This procedure is described in Fogel [4] to evolve neural networks in mathematical games, and in Fogel and Simpson [5] to evolve clusters based on fuzzy membership functions. Information criteria can be applied to design evolutionary networks in a similar manner to the construction of models in system identification [6]. The self-design process is almost automatic; unlike traditional neural network paradigms that require the active participation of the user as part of the learning algorithm, an evolutionary neural network can adapt to unexpected feature inputs on its own, or with little operator intervention. The resulting system is more robust than traditional approaches in symbolic artificial intelligence, and is capable of machine learning.

When simulated evolution has been used to train neural networks (e.g., [7]), the results have been superior to those from other methods. Porto et al. [8] compared back propagation, simulated annealing, and evolutionary programming for training a fixed network topology to classify active sonar returns. The results indicated that stochastic search techniques such as annealing and evolution consistently outperformed back propagation, yet can be executed more rapidly on an appropriately configured parallel processing computer. After sufficient computational effort, the most successful network can be put into practice. But the evolutionary process can be continued during application, so as to provide iterative improvements on the basis of newly acquired exemplars. The procedure is efficient because it can use the entire current population of networks as initial solutions to accommodate each newly acquired datum. There is no need to restart the search procedure in the face of new data, in contrast with many classic search algorithms, such as dynamic programming.

Designing neural networks through simulated evolution follows an iterative procedure:

1. A specific class of neural networks is selected. The number of input nodes corresponds to the amount of input data to be analyzed. The

number of classes of concern (i.e., the number of output classification types of interest) determines the number of output nodes.

2. Exemplar data is selected for training.
3. A population of *P* complete networks is selected at random. A network incorporates the number of hidden layers, the number of nodes in each of these layers, the weighted connections between all nodes in a feedforward or other design, and all of the bias terms associated with each node. Reasonable initial bounds must be selected for the size of the networks, based on the available computer architecture and memory.
4. Each of these "parent" networks is evaluated on the exemplar data. A payoff function is used to assess the worth of each network. A typical objective function is the mean-squared error between the target output and the actual output summed over all output nodes; this technique is often chosen because it simplifies calculations in the back propagation training algorithm As evolutionary computation does not rely on similar calculations, any arbitrary payoff function can be incorporated into the process and can be made to reflect the operational worth of various correct and incorrect classifications. Information criteria such as Akaike's information criterion (AIC) [9] or the minimum description length principle [5] provide mathematical justification for assessing the worth of each solution, based on its classification error and the required degrees of freedom.
5. "Offspring" are created from these parent networks through random mutation. Simultaneous variation is applied to the number of layers and nodes, and to the values for the associated parameters (e.g., weights and biases of a multi-layer perceptron, weights, biases, means and standard deviations of a radial basis function network). A probability distribution function is used to determine the likelihood of selecting combinations of these variations. The probability distribution can be preselected *a priori* by the operator or can be made to evolve along with the network, providing for nearly completely autonomous evolution [4].
6. The offspring networks are scored in a similar manner as their parents.
7. A probabilistic round-robin competition is conducted to determine the relative worth of each proposed network. Pairs of networks are

selected at random. The network with superior performance is assigned a "win." Competitions are run to a preselected limit. Those networks with the most wins are selected to become parents for the next generation. In this manner, solutions that are far superior to their competitors have a corresponding high probability of being selected. The converse is also true. This function helps prevent stagnation at local optima by providing a parallel biased random walk.

8. The process iterates by returning to step (5).

3. Breast Cancer

Carcinoma of the breast is second only to lung cancer as a tumor-related cause of death in women. There are upwards of 180,000 new cases annually in the United States, with approximately 45,000 deaths [10]. These statistics are little changed over the past decade [11]. Breast cancer begins as a focal curable disease, but it is usually not identifiable by palpation at this stage. Therefore, mammography remains the most-effective current screening method. Strax [12] suggested that mortality from breast cancer could be decreased by 25 percent if all women in the appropriate age groups were screened regularly.

3.1 Mammographic Screening

Interest in computer-aided decision making in screening for breast cancer has a long history, with significant efforts made to apply computational intelligence techniques to this problem over the past decade. In particular neural networks have been used for pattern recognition on features associated with the disease. Wu et al. [13] used 43 prescribed mammographic features related to abnormal density, microcalcification, parenchymal distortion, skin thickening, correlation with clinical findings, etc. Data were extracted from 133 textbook cases in [14]. For each mammogram, each selected feature was rated by an experienced mammographer on a ten-scale, with this serving as the vector input to a multilayer feedforward perceptron. The neural network had 10 hidden units and 1 output unit, which was trained to yield a value of 0.0 for a benign case and 1.0 for a malignancy. The neural network was trained using the gradient backpropagation method. Wu et al. [13] pruned the feature set successfully and demonstrated a neural network that could outperform attending radiologists and residents in assessing patterns of mammographic image features associated with benign and malignant lesions. There was no demonstrable difference between the performance of the neural network and the experienced mammographer who rated the image features.

Floyd et al. [15] applied a similar approach, working with only eight input parameters:

1. Mass size.
2. Mass margin.
3. Asymmetric density.
4. Architectural distortion.
5. Calcification number.
6. Calcification morphology
7. Calcification density.
8. Calcification distribution.

A total of 260 cases were used in training and testing via a jackknife statistical procedure. The results of "significant training" showed that it was possible to identify all cancerous cases correctly (92 malignancies) while also correctly rejecting 38 of 168 benign cases.

These and other efforts (e.g., [16]) relied on gradient backpropagation optimization of a fixed neural architecture. As indicated above, a potential advantage of an evolutionary approach for optimizing the performance of a neural network in these cases was recognized. Fogel et al. [17] reported results from a study using this combination of approaches.

Data were collected by assessing film screen mammograms in light of 12 radiographic features as determined by a domain expert (Eugene Wasson). The features selected paralleled those offered in [15], with the inclusion of patient age. These features were assessed in 112 cases of suspicious breast masses, all of which were examined subsequently in open surgical biopsy with the associated pathology indicating whether or not a malignant condition had been found. In all, 63 cases were associated with a biopsy-proven malignancy, while the other 49 cases were negative by biopsy.

These data were processed using a feedforward neural network with only two hidden nodes and a single output node, resulting in 33 adjustable weights. Evolutionary optimization was used to train the neural network in a leave-one-out cross validation where each sample pattern was held out for testing and then reinserted in a series of 112 separate training procedures. The evolutionary algorithm relied on a population of 250 networks, initialized randomly with weights distributed uniformly over [-0.5, 0.5]. Self-adaptation of the weights was employed using the standard approach of incorporating a vector σ along with each weight set, which is itself subject to variation. Cauchy mutations were applied to all 33 weights, simultaneously. See Fogel [17] for further details. Each weight set was evaluated on how well it allowed the neural network to classify the 111 training samples in each run in terms of the sum of the squared error (the targets were 1.0 for malignancy and 0.0 for a benign case). Selection eliminated the lower half of the population of neural networks at each generation, with 200 generations being executed during the experiment.

A typical rate of optimization in any single trial is shown in Figure 1. The effectiveness of the procedure was assessed using receiver operating characteristic (ROC) analysis, where the probability of detecting a malignancy is traded off as a function of the likelihood of a false positive. A typical ROC curve from 16 separate trials conducted is offered in Figure 2. The area under the curve, denoted as A_z, provides a useful measure for assessing the performance of the system. The mean A_z was 0.8982 with a standard error of 0.0098 over 16 trials. The best neural network achieved an A_z of 0.9345.

Fogel et al. [17] observed that the average performance of the evolved neural networks in terms of A_z was comparable to the above-mentioned prior studies, which also used mammographic features interpreted by a radiologist. For example, the results in [18], which relied on 18 input features (both radiographic and clinical) used 10 hidden nodes to achieve a specificity of 0.62 and a sensitivity of 0.95. (Radiologists achieved only 0.3 specificity on the same data.) The evolved neural networks in Fogel et al. [17] achieved a mean specificity of 0.6187 ± 0.0285 at 0.95 sensitivity. Although this result is comparable to that in [18], the evolved neural networks were more parsimonious (about an order of magnitude fewer degrees of freedom), and may therefore offer greater generalizability.

Fogel et al. [19] further explored the evolution of neural networks for screening features from mammograms. Attention was given to 216 cases with suspicious characteristics. Wasson quantified up to 12 radiographic features for each case, along with patient age. The existence or absence of malignancy was confirmed in each case via open surgical biopsy (111 malignant, 105 benign). Artificial neural networks of various complexity were trained using evolutionary algorithms in cross validation and the resulting mean area under the receiver operating characteristic curve was 0.9196 ± 0.0040, with a mean specificity of 0.6269 ± 0.0272 at 0.95 sensitivity. The results when microcalcifications were included were not quite as good (A_Z = 0.8464), however, neural networks with only two hidden nodes per formed as well as more complex neural networks and better than neural networks with only one hidden node.

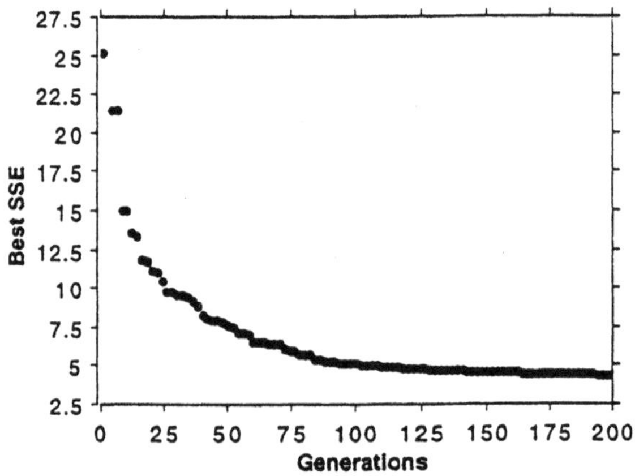

Figure 1. Typical optimization performance using simulated evolution to train an artificial neural network. The graph depicts the sum of squared error (SSE) of the best network in the population as a function of the number of generations. Training was performed over 111 patterns, with one pattern held out for testing in cross validation. The sufficiency of the number of generations is inidcated as the learning curve approaches an asymptote.

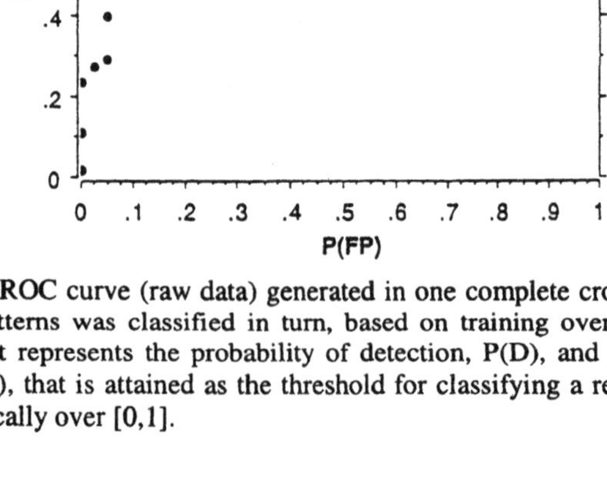

Figure 2. A typical ROC curve (raw data) generated in one complete cross validation where each of the 112 patterns was classified in turn, based on training over the remaining 111 patterns. Each point represents the probability of detection, P(D), and the probability of a false positive, P(FP), that is attained as the threshold for classifying a result as malignant is increased systematically over [0,1].

Figure 3 shows a typical ROC curve generated for the evolved neural classifier in one complete cross validation where each of 158 patterns was classified in turn, based on training over the remaining 157 patterns. Each point represents the probability of detection, P(D), and probability of false positive, P(FP), that is attained as the threshold for classifying a result as malignant is increased systematically over [0,1]. Using a polynomial spline, the area under the ROC curve depicted here is estimated at 0.9147.

Figure 4 shows the mean and one-standard-error bars on the output of the best-evolved neural networks across all 158 classifications in the 16th (final) trial. There is a good separation between the two output classes, indicating that the neural classifiers are capable of discriminating between the benign and malignant cases.

Figure 5 compares the ROC curves for neural networks with hidden nodes varying in number from one to five on the complete set of 216 cases. The curves are depicted as connected graphs for ease of comparison. The ROC curve for the classifier with only one hidden node provides the lowest area under the curve. Including the capability for multiple nonlinear discrimination (i.e., using two hidden nodes) improves performance; however, including additional degrees of freedom in the form of more hidden nodes does not improve on that performance attained by using only two hidden nodes. This provides justification that the two-hidden node architecture is a reasonable choice for the available data.

Figure 6 shows a sampled neural network's output recorded across the range of values of mass size and patient age when all other inputs were held constant at their respective means. The network's output is seen to increase as a sigmoid function of patient age, as anticipated, but appears independent of mass size. This could reflect earlier detection of smaller cancers and the presence of large benign masses (fibro adenomas) in the available data. This "negative result" suggests the possibility of pruning the inputs to a minimal set thereby focusing attention on the truly crucial features while providing an even greater chance of explaining the evolved decision rules that are captured by the neural networks, leading to a greater acceptance by physicians.

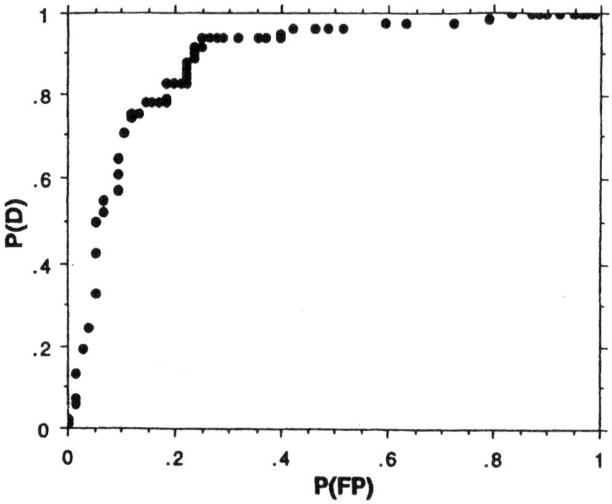

Figure 3. A typical ROC curve (raw data) generated for the artificial neural network classifier in one complete cross validation where each of 158 patterns was classified in turn, based on training over the remaining 157 patterns. Each point represents the probability of detection, P(D), and the probability of false positive, P(FP), that is attained as the threshold for classifying a result as malignant is increased systematically over [0,1]. Using a polynomial spline, the area under the ROC curve is estimated at 0.9147, which was slightly below the mean of 0.9196 observed in16 independent trials.

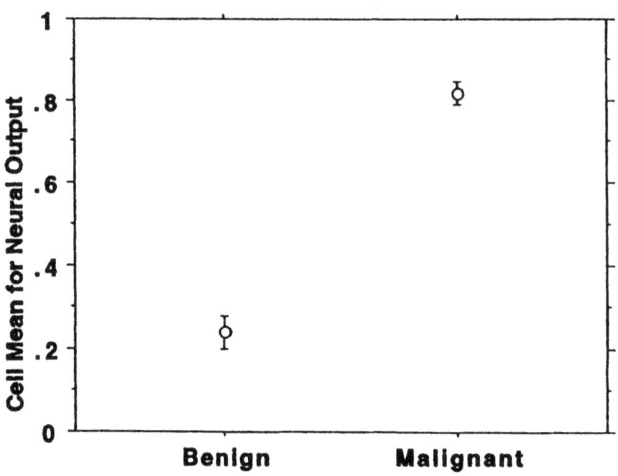

Figure 4. The mean and one-standard-error bars on the output of the best-evolved neural networks across all 158 classifications in the sixteenth trial. There is a good separation between the two output classes, indicating that the neural classifiers are capable of discriminating between the benign and malignant classes.

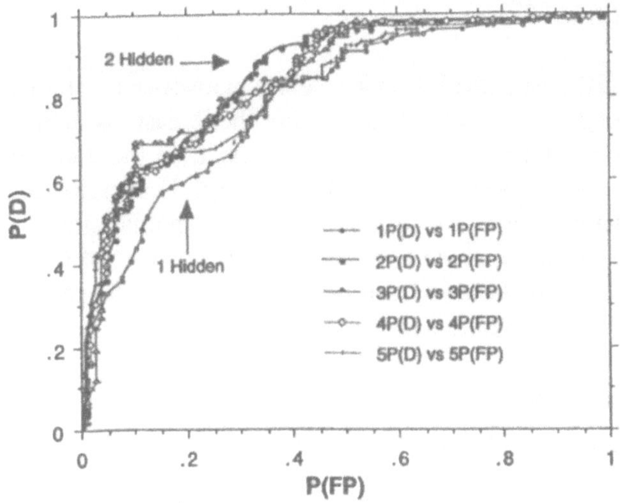

Figure 5. Comparing ROC curves for neural networks with hidden nodes varying in number from one to five. The curves are depicted as connected graphs for ease of comparison. Two hidden nodes provides significant performance improvement over one hidden node, but more than two hidden nodes does not appear to offer increased performance.

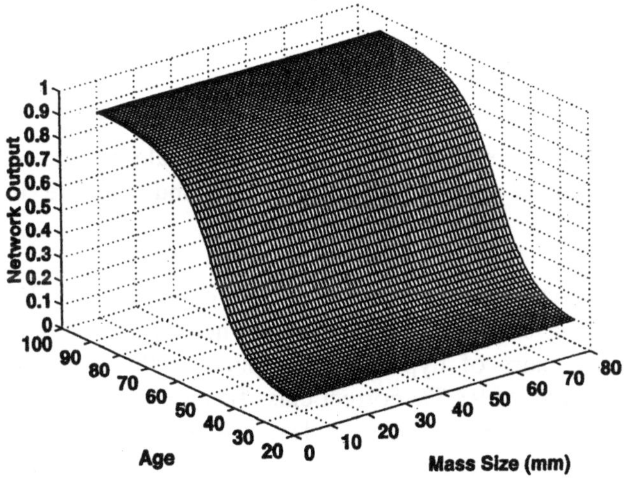

Figure 6. A sampled neural network's output recorded across the range of values of mass size and patient age when all other inputs were held constant at their respective means. The network's output is seen to increase as a sigmoid function of patient age, as anticipated, but appears independent of mass size. This could reflect earlier detection of smaller cancers and the presence of large benign masses (fibro adenomas) in the available data.

3.2 *Classifying Breast Cells*

Fogel et al. [20] used evolutionary neural networks to analyze a related problem of classifying cells taken from the breast as being benign or malignant. Considerable analysis has been made on the data offered by O. Mangasarian and collected by W.H. Wolberg, University of Wisconsin Hospitals (available on ics.uci.edu). Previous publications with earlier versions of these data include [21, 22]. The data consist of 699 instances of parameterized histopathology from breast biopsies. Each specimen has been assessed with regard to nine parameters:

1. Clump thickness.
2. Uniformity of cell size.
3. Uniformity of cell shape.
4. Marginal adhesion.
5. Single epithelial cell size.
6. Bare nuclei.
7. Bland chromatin.
8. Normal nucleoli.
9. Mitosis.

Each parameter has been rated on an integer 10-scale. Of the 699 patterns, 458 are indicated as known benign (65.5%) and 241 are indicated as malignant (34.5%).

Based on data supplied in January, 1989, consisting of 369 samples, Wolberg and Mangasarian [23] showed that when half of the data were used for training, two pairs of hyperplanes applied to the parameterized data could achieve a level of 93.5% correct classification, and three pairs of such parallel hyperplanes could achieve a level of 95.9% correct. Zhang [24] operating on the same data, showed that a 1-nearest neighbor technique could classify 93.7% of the data correctly (trained on 200 instances).

Fogel et al. [20] used an evolutionary approach to training a small neural network for the classification problem. Sixteen of the 699 data had missing values and were removed, leaving 683 data. The first 400 data were chosen as the training set while the remaining 283 were held out for testing. Two experimental designs were conducted. The first experiment consisted for five trials with a 9-2-1 multilayer perceptron (nine inputs, two hidden nodes, one output node) and five trials with a 9-9-1 multilayer perceptron. A population of 500 networks was evolved over 400 generations in each trial. Each parent was evaluated in terms of its total squared error (deviation from output to target value assigned to malignant or benign case). One offspring network was created from each parent at each generation. Each weight of each parent was mutated by adding a zero-mean Gaussian random variable. The standard deviation of each mutation was set proportional to the parent's total squared error (i.e., the poorer the performance, the larger the mutation applied). After all of the offspring networks were scored, the networks underwent a tournament with 10 competitions (typical tournament selection). The 500 networks with the most wins were selected to become parents for the next generation. The best network at the end of each trial was used to classify the test set. The second experiment extended the first to a sample size of 16 trials with a 9-2-1 multilayer perceptron based on results observed in the first experiment.

The results of the first experiment are indicated in Table 1. After 400 generation in each of five trials, the average mean squared error on the training set was about 0.13 with the 9-9-1 network and about 0.11 with the 9-2-1 network. The test set performance appeared to favor the smaller network configuration (see Table 1 discussion). The results of the second experiment are indicated in Table 2. The test set performance was statistically significantly superior to the best performance reported in the literature on related data from the archive.

These results, although preliminary, indicated that a reasonably high level of performance could be achieved by small neural networks operating on histopathologic features from breast biopsies. The classification results presented on these data surpassed the best results reported before Fogel et al. [19] and routinely achieved greater than 97% correct. No explicit attempts were made to force the classification results to favor false positives, but the networks tended to err in that direction.

4. Basic Research Comparing Alternative Variation Operators in Evolving Neural Networks

Evolutionary algorithms are often used for optimization in an attempt to find the best solution(s) to a problem that is defined in terms of an evaluation function. In this regard, these algorithms apply a version of stochastic search over a state space of potential candidate solutions that is defined by a representation and a collection of variation operators. The search can be captured in the difference equation:

$$\boldsymbol{x}[t+1] = s(v(\boldsymbol{x}[t]))$$

where $\boldsymbol{x}[t]$ is the population at time t under representation $\boldsymbol{x}$, v is the variation operator(s), and s is the selection operator. The stochastic components of this update equation involve the use of random variation (v) and potentially probabilistic selection (s), and random initialization of $\boldsymbol{x}[0]$. The choices that are made for representation, variation, and selection impact the improvement that is observed on average between the elements of $\boldsymbol{x}[t]$ and $\boldsymbol{x}[t+1]$.

Trial No.	MSE on training	Test Set errors and % correct
9 hidden nodes		
1	0.12816	8 errors (6/2) 97.1%
2	0.12484	7 errors (6/1) 97.5%
3	0.12899	6 errors (5/1) 97.9%
4	0.12605	7 errors (6/1) 97.5%
5	0.12593	7 errors (5/2) 97.5%
2 hidden nodes		
1	0.10409	6 errors (5/1) 97.9%
2	0.11885	4 errors (3/1) 98.6%
3	0.10389	4 errors (3/1) 98.6%
4	0.10952	4 errors (3/1) 98.6%
5	0.10397	5 errors (5/0) 98.2%

Table 1. Performance in the first experiment in which 500 networks were evolved over 400 generations using both nine and two hidden nodes. The more parsimonious networks with only two hidden nodes generated superior training (400 samples) and test performance (283 samples). The additional dimensions (100 versus 23) may have slowed training on the 9-9-1 networks. A t-test on the two-sample mean difference in testing is significant ($P < 0.002$) although the assumption of normality required for the test can be questioned. The numbers in parentheses indicate the number of false positives and false negatives, respectively. The percentages indicate the percent correct. MSE refers to mean squared error.

Average MSE on training	0.10841 (0.00461)
Average % correct on test set	98.05 (0.46476)
Mean number of false positives	4.25 (1.06458)
Mean number of false negatives	1.25 (0.77460)

Table 2. Results of the second experiment comprising 16 independent trials with the 9-2-1 network. The test set performance (283 samples) is statistically significantly superior to the

best performance indicated in the literature with related data (95.9%) at $P < 0.05$ under nonparametric testing based on the Chebyshev inequality and $P << 0.001$ under a t-test. The numbers in parentheses indicate the sample standard deviations. MSE refers to mean squared error.

Many early suggestions for how to best choose the representation, variation, and selection operators have proved less than useful. For example the recommendations for binary encodings, crossover, and proportional selection found in [25] have all been largely eschewed based on practical and theoretical results. It is now known that no representation is optimal for any single problem (Fogel and Ghozeil [26]), that no algorithm that does not resample points can have an advantage over any others across all problems (Wolpert and Macready [27]), and the claim [25] that proportional selection "minimizes expected losses" in sampling schemata is not correct [28, 29].

When considering an arbitrary function to optimize, it may be necessary to rely on empirical observations to determine the appropriate settings for variation parameters or operators. When a variation operator can be defined in terms of a scaling parameter, it becomes possible to generate a "fitness distribution" that indicates the mean fitness of an offspring from a particular parent (or collection of parents) as a function of the parameter [30-33]. This information can often indicate whether or not a particular operator is serving to accelerate or decelerate the evolutionary search for improved solutions. Furthermore, this information can be used to compare operators.

For example, Nordin and Banzhaf [34] quantified the fitness change that occurred after crossover was applied to machine code expressions that served as regression equations. Figure 7 shows the percentage of fitness change (positive is an improvement) as a histogram over 35 successive generations. Note that the most common effects were either no change at all, meaning that the offspring's fitness was the same as that of its parents, or a 100 percent or larger decrease in fitness. The construction of superior solutions from "building blocks" of useful code appeared absent. Moreover, the ineffectuality of crossover increased with increasing generations: More and more, here crossover did less and less. This effect probably occurred because crossover mostly recombined sections of different parents that had no effect on the fitness of a solution, and these sections became more frequent as evolution proceeded.

Fitness distributions can also be used to assess the appropriateness of settings for operators when evolving neural networks. Experiments have been conducted to identify the sensitivity of Gaussian mutation when optimizing a simple neural network for classifying the well-known exclusive-or (XOR) problem [35]. The results indicated that fitness distributions can not only indicate appropriate settings for operators but can also highlight changes in the local response surface and indicate the presence of multiple local minima. Here, the experiments are extended to compare the effects of one-point, two-point, and uniform crossover in optimizing the classification performance of a small neural network. The results indicate little difference in effect between the three operators, but a significant change in their potential utility as a function of the number of generations.

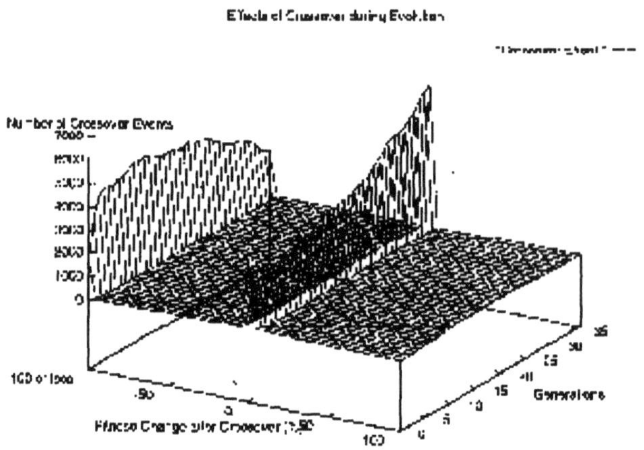

Figure 7. The fitness change after crossover on the problems evolving machine code expressions from [34]. The vast majority of crossovers either resulted in no change or a large decrease in offspring fitness.

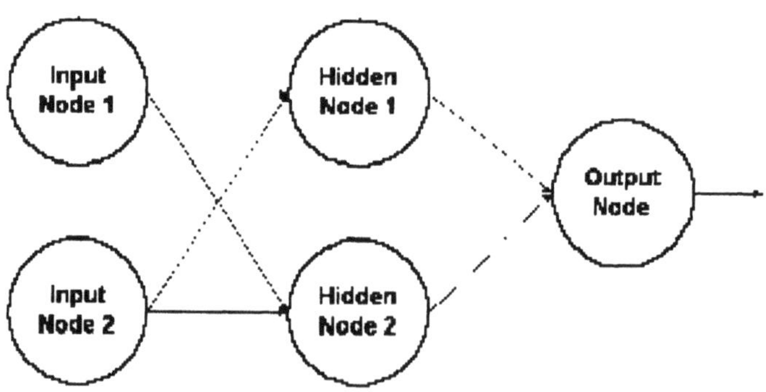

Figure 8. The artificial neural network topology used for classifying instances in the XOR problem.

4.1 Background on Analysis

The XOR task is well known [36]. The four possible inputs of (0,0), (0,1), (1,0), and (1,1) must be mapped to outputs of 0, 1, 1, and 0, respectively. A network comprising two input nodes, two hidden nodes, and a single output node was chosen (Figure 8), where each of the hidden and output nodes used a logistic (sigmoid) filter of the form $(1 + e^{-x})^{-1}$, where x is the sum of an adjustable threshold term and the dot product of the incoming activations and their associated weights. This architecture is known to be able to compute the XOR mapping [36].

In [35], the nine weights of the neural network were evolved using the following procedure. A population of 50 neural networks was initialized with each weight (and threshold, hereafter also termed a weight) sampled from a uniform distribution over the range [−0.5, 0.5]. Each network was evaluated in terms of the sum of the squared error between the target output and the actual output taken across all four input patterns. The 50 initial parents each generated a single offspring by varying all weights simultaneously in accordance with a zero-mean Gaussian random variable with a fixed standard deviation of 0.2. That is:

$$w'_{ij} = w_{ij} + N(0, 0.2)$$

where w'_{ij} is the jth weight of the ith parent and $N(0, \sigma)$ is a zero-mean normal random variable with standard deviation σ. It is recognized that this procedure is not optimal for converging on suitable neural classifiers, as this would likely require an adaptive step size in each dimension (weight). For the experimental purposes here, however, the procedure was adequate in that it was sufficient to generate an optimal neural network and the networks that were generated at specific generation intervals could be examined. After generating and evaluating each set of 50 offspring, the 50 best networks from among all 100 in the population were selected as parents of the next generation and the process was iterated for 200 generations.

Attention was then turned to the best-evolved networks at generations 0, 50, 100, 150, and 200. Each such network was mutated using zero-mean Gaussian mutations with a standard deviation that varied from 0.01 to 2.0 at steps of 0.01. For each step size, 10,000 offspring were generated and the degree of improvement over the respective parent was calculated as the decrease in error of the offspring relative to the parent at that particular iteration. If the improvement was negative (i.e., the offspring was worse than its parent), it was taken to be zero, this in accordance with a so-called "plus strategy" selection where offspring only replace parents if they are superior. The average improvement for each step size was termed the expected improvement (note that it is an estimate of this parameter). Taken across the range of step sizes, each set of expected improvement values generates a fitness distribution. Based on the results obtained, further experimentation was conducted by extending the range of the standard deviation applied at generation 0 to [0.1, 14] by steps of 0.1, averaged over 200,000 samples at each setting, and by perturbing the best-evolved networks at generations 12, 18, 20, 22, 25, 37, 43, and 46 with zero-mean Gaussian noise sampled over [0.1, 10] in order to determine how the fitness distribution changed from its initial characteristics to those found in generations 50 and beyond.

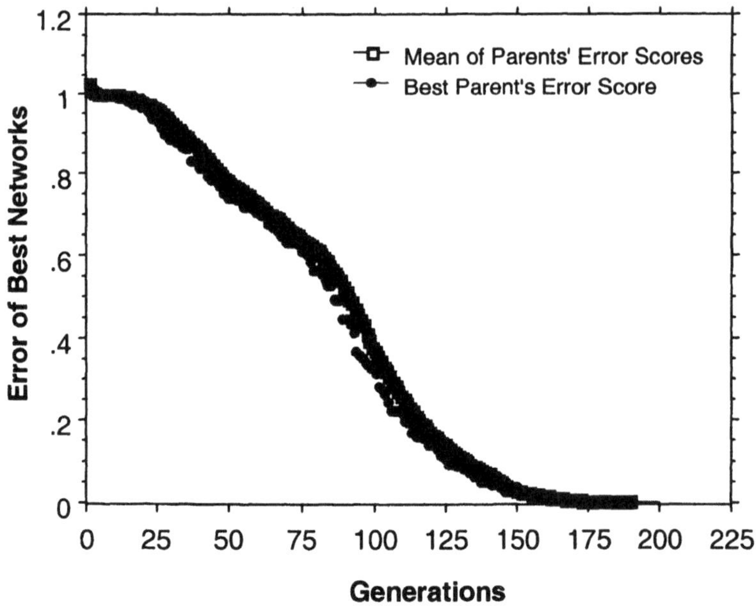

Figure 9. The optimization of the best neural network and the mean of all parent networks at each generation under training. The evolutionary adaptation was able to find neural networks that solve the XOR problem within 200 generations. Evolution was seen to advance at three different rates: (1) an initial slow rate, (2) a faster intermediate rate, (3) a final apparently exponential rate as the process approached neural networks that can solve the problem.

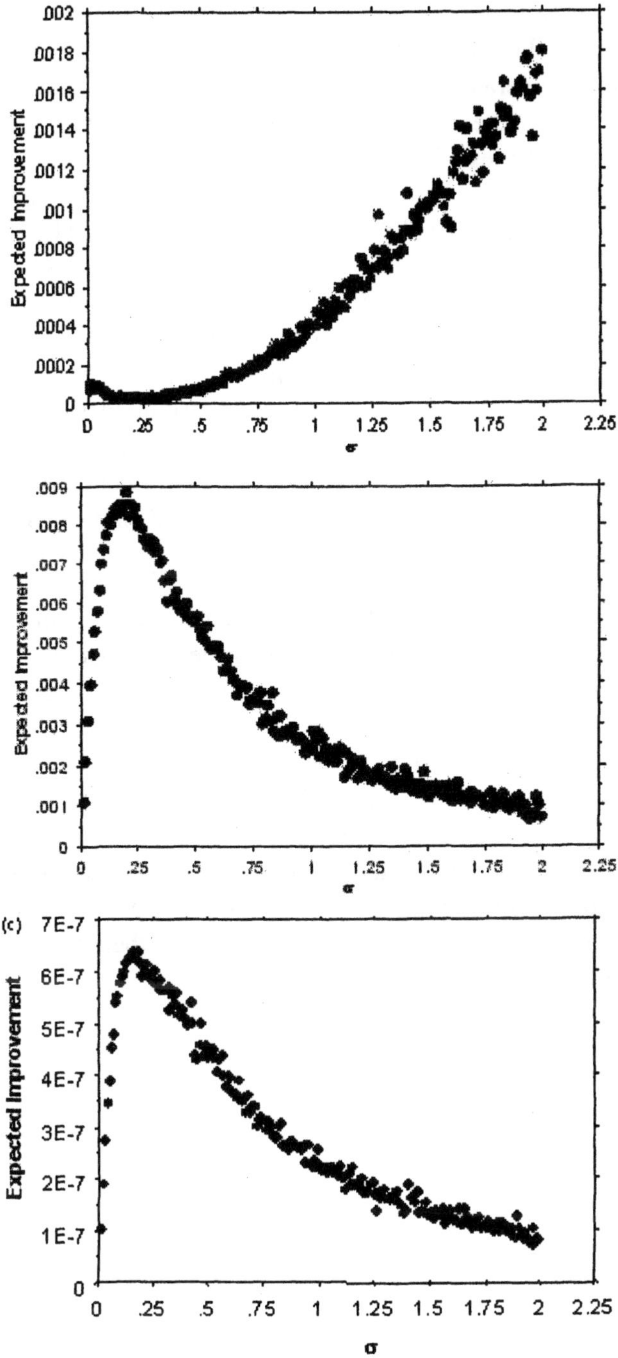

Figure 10. Three fitness distribution graphs for generations (a) 0 (above) and (b) 50, and (c) 200.

The error of the best-evolved network decreased as a function of the number of generations as shown in Figure 9. The curve indicates that suitable XOR classifiers were generated by the 100th generation, and that nearly perfect mapping functions were evolved within the 200 generations allowed. Figure 10(a-c) shows the fitness distributions obtained when focusing on the best-evolved neural networks from generations 0, 50, and 200, respectively. The distributions changed gradually from generation 50 to 200, but showed a significant dissimilarity between the initial fitness distribution and those obtained in later generations. Whereas the optimal value of the standard deviation at generations 50 and beyond was approximately $\sigma = 0.2$, this is curiously close to the worst possible setting on generation 0, given the range from [0.01, 2]. Figure 11 shows the fitness distribution for settings of σ at generation 0 when the range of step sizes was extended to [0.1, 14], sampled at intervals of 0.1. The maximum expected improvement was observed at a value of approximately $\sigma = 5.8$, with the expected improvement declining gradually after that.

It was of interest to determine the characteristics of the transition between this fitness distribution and that observed for generations 50, 100, 150, and 200, which are mutually similar. Figure 12 shows the fitness distributions for applying Gaussian variation to the best-evolved neural networks from generations 12, 18, 20, 22, and 25. The figure indicates that the best setting for σ did not move gradually from 5.8 at generation 0 to 0.2 at generation 50. Instead, a separate local maximum setting of σ arose by the 12th generation at 0.3. The expected improvement for this local maximum increased with increasing generations until it overtook the previous maximum by generation 22. The best setting for σ then moved from 0.3 to 0.2 during generations 43-50 (Figure 13).

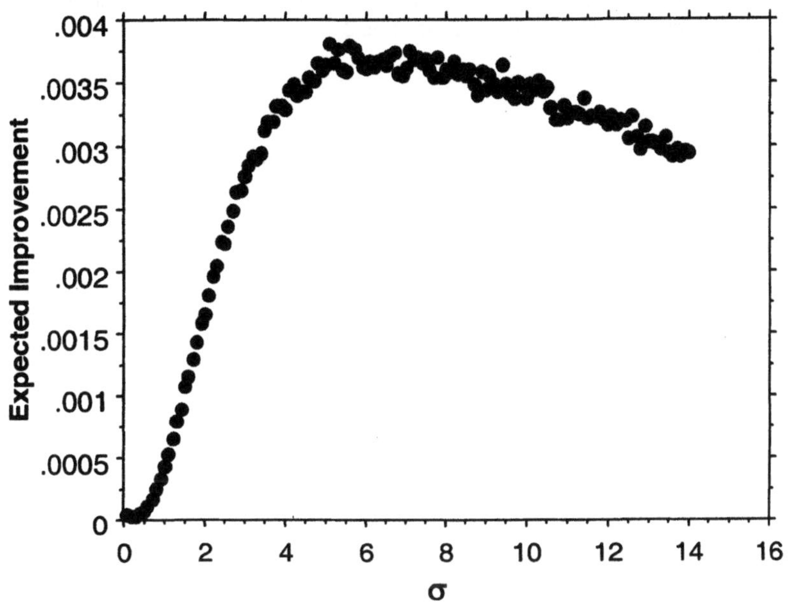

Figure 11. The fitness distribution at generation 0 when σ is varied over [0,14]. The graph indicates a local minimum expected improvement at small values of σ and a maximum expected improvement at about σ = 5.8.

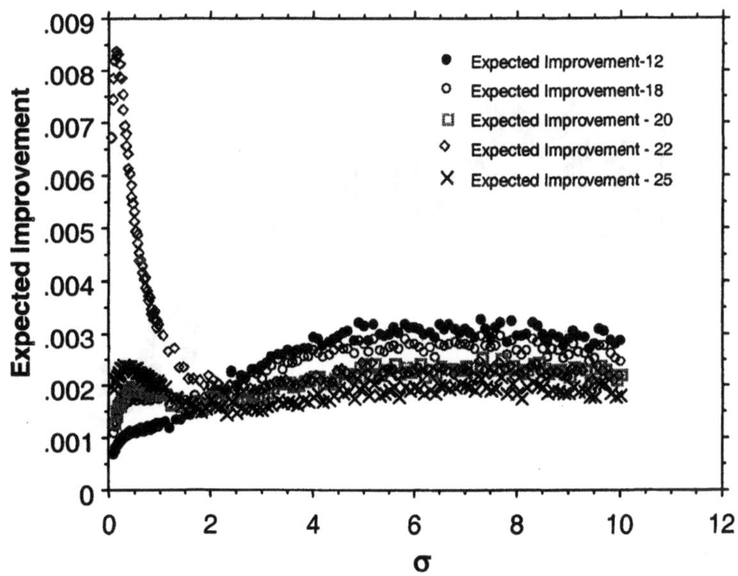

Figure 12. The fitness distributions for generations 12, 18, 20, 22, and 25. The sequence of distributions show a change from a maximum expected improvement of about σ = 5.8 to σ = 0.2, but this change does not occur gradually. Instead, the best setting for σ jumps from high to low values as the fitness distribution reflects the apparent multimodality of the response

surface around the best parent found from generations 12-25. Note that the expected improvement increases dramatically for generation 22.

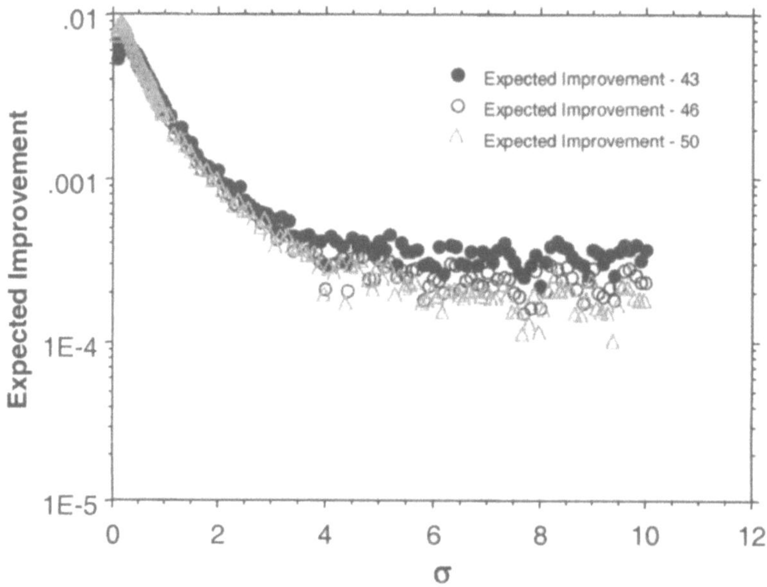

Figure 13. The fitness distributions for generations 43, 46, and 50. The distributions are all similar and all indicate that the best choice of σ is about 0.2.

(a)

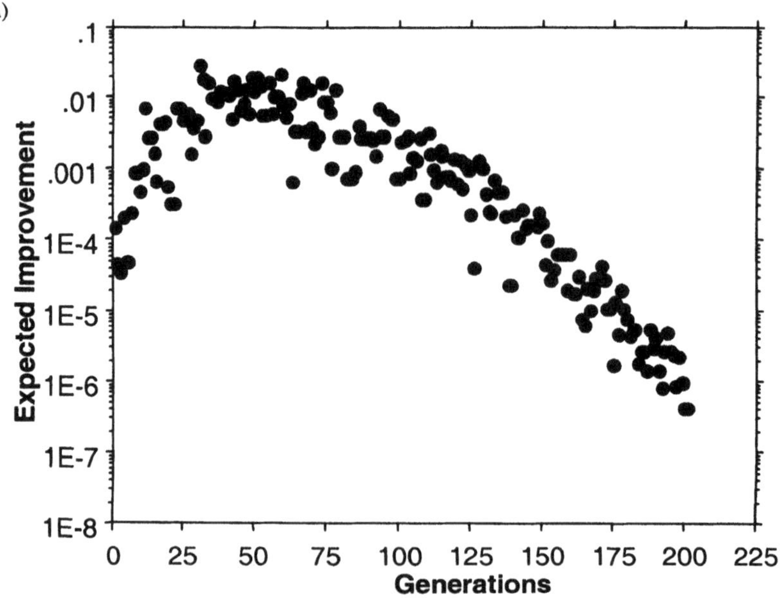

(b)

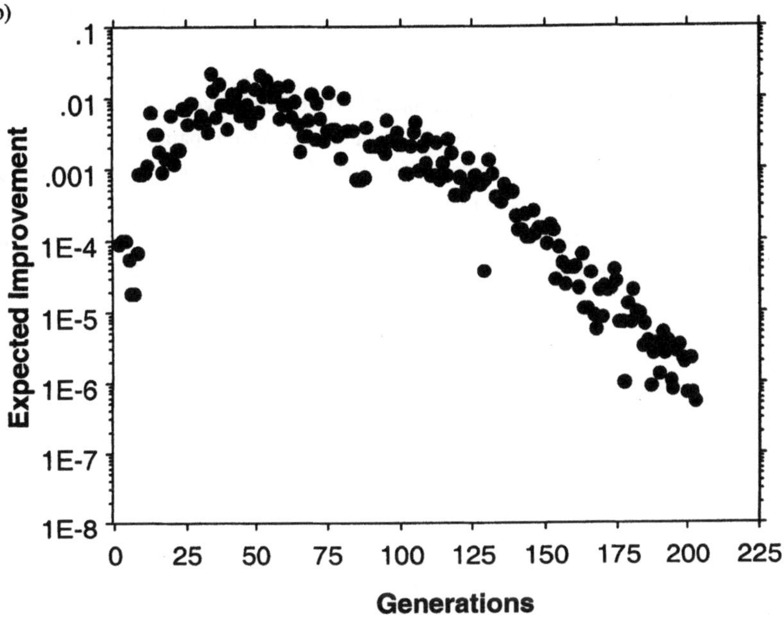

(c)

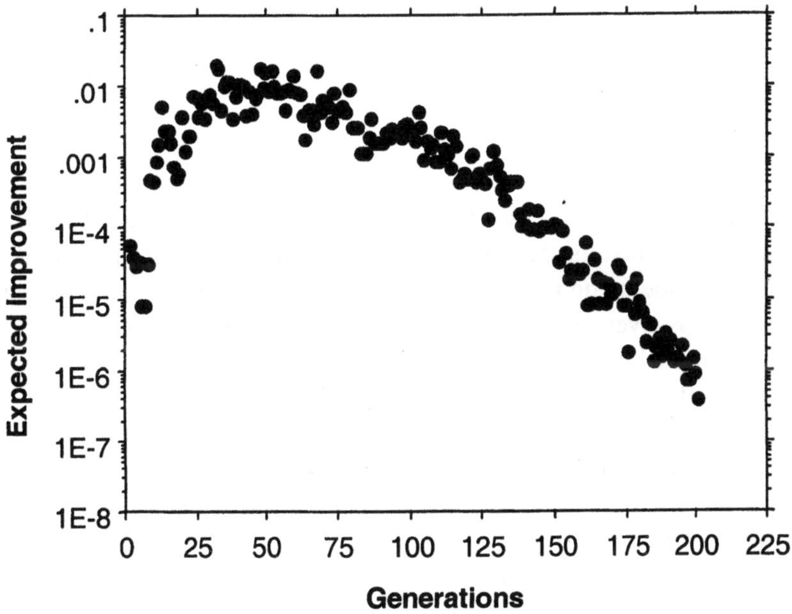

Figure 14. The fitness distributions of (a) one-point, (b) two-point, and (c) uniform crossover as a function of the number of generations when applied to the best-evolved networks at that generation. The results indicate that all three crossover operators provide similar utility. The

peak utility comes at about the 50th generation and then falls exponentially in subsequent generations.

Table 3. Comparison of expected improvement between crossover operators and Gaussian mutation constrained to varying a single standard deviation parameter. Note that the calculation of the expected improvement differs slightly between the case of a single parent and the case of two parents. Gaussian mutation is seen to outperform crossover at the initial generation. Crossover outperforms Gaussian mutation at the 50th generation. Beyond the 100th generation, both methods of variation generate similar expected improvement.

Generation	1-Point Crossover	2-Point Crossover	Uniform Crossover	Gaussian Mutation
0	1.4216e-4	9.3698e-5	5.3828e-5	3.8001e-3
50	1.8846e-2	1.7556e-2	1.6799e-2	8.5755e-3
100	2.3210e-3	2.0607e-3	1.6450e-3	1.4039e-3
150	4.4840e-5	2.8316e-5	3.1290e-5	2.3103e-5
200	4.1200e-7	5.4400e-7	3.7800e-7	6.4000e-7

4.2 *Method and Results*

Experiments were conducted to study the effects of applying one-point, two-point, and uniform crossover to the best-evolved networks at each generation. The results are then compared and also contrasted with the gains that can be made using Gaussian mutation. The standard instances of each of the three crossover operators were used in which crossover points were chosen without bias for the one- or two-point crossover operators, and the probability of selecting from either parent was 0.5 under uniform crossover. In each case, the two best-evolved networks in the population were chosen and all possible one- and two-point crossovers were determined. For uniform crossover, 10,000 samples were generated. The expected improvement was calculated as the improvement of the average of the best two networks taken from the parents and offspring (2+2) to the average score of the two parents. Figure 14 shows the expected improvement as a function of the generation number. All three crossover operators demonstrate approximately equal performance. Table 3 provides data for comparing the utility of recombination to Gaussian mutation based on a single standard deviation. The results indicate that simple mutation outperforms recombination in the early stages of the search, whereas recombination outperforms simple mutation at about the 50th generation. Beyond this, both forms of variation generate roughly equal performance.

4.3 Discussion

The results from these simple experiments indicate some fundamental aspects about the application of variation operators in evolutionary algorithms. The primary insight is that the effectiveness of a variation operator can (and likely will) change over time as a function of the current members of the population. Previous efforts in [35] showed this was true for the simple Gaussian mutation operator. The current experiments demonstrate that this also holds for crossover operators. The traditional practice of setting operator probabilities at constant values, as is often observed even now in genetic algorithms (e.g., probability of crossover = 0.6, probability of mutation = 0.01), is quite limiting and may even prevent the successful discovery of suitable solutions. The information contained in the fitness distributions taken over a series of generations can indicate apt means for varying the settings of the search operators, or choosing between them.

The fitness distributions for all three crossover operators appeared very similar. This evidence runs counter to the so-called "building block hypothesis," which suggests that gains are made in evolutionary algorithms by recombining schemata that are associated with above-average fitness. Here, uniform crossover, which is maximally disruptive of schemata when the probability of choosing either parent for each component is 0.5, performed on par with one-point and two-point crossover. Evidently, the ability to improve performance over existing parents did not depend on recombining schemata in these cases. This is consistent with other studies of optimizing neural networks that have suggested an inability to perform effective recombination on neural networks [37, 38].

When comparing the performance of Gaussian mutation against the three forms of recombination, there was essentially no difference in the utility of any of these operators after the 100th generation. Amongst these four options, the choice of variation operator was effectively irrelevant. This presents data consistent with the general perspective that there is nothing inherently unique about any particular form of evolutionary algorithm that would necessarily distinguish its performance in contrast with another. There appears to be little reason to discriminate between types of evolutionary algorithms (e.g., based on the use of recombination) when performance can be, for all intents and purposes, identical.

The previous study [35] constrained the effectiveness of Gaussian mutation by requiring a single standard deviation that controls the step size in all nine dimensions. Allowing for independent step size variation in each dimension should enable more effective optimization. It remains to be seen if a multivariate Gaussian mutation operator with independent step sizes can generate orders of magnitude improvements over one that relies on a single step size, or over any of the crossover operators explored here.

5. Remarks on Walter Karplus' Influence on Computational Intelligence

In closing this paper, I would like to offer a few brief remarks on Walter Karplus' influence on computational intelligence, and more specifically on me. I came to know Walter in the early 1990s as part of his efforts in the IEEE Neural Networks Council. Walter impressed me as a person with broad perspective, able to see the interdisciplinary nature of the problems of computational intelligence, and the links between neural, fuzzy, and evolutionary computation, as well as associated technologies, such as virtual reality. His longstanding vision for the Neural Networks Council was for it to become a full-fledged society in the IEEE, a vision that became a reality shortly after his passing. I regret that he was not able to experience this important occasion.

I have no regrets, however, about the encouragement that Walter gave me, and many others within the computational intelligence community. In particular, in 1995 at the joint meeting of the IEEE neural network and evolutionary computation conferences in Perth, Australia, Walter took me aside and asked me to begin the process for creating the *IEEE Transactions on Evolutionary Computation.* I found his faith in my ability to lead this effort both reassuring and humbling, and indeed Walter was a humble individual himself. In significant proportion, the success we have enjoyed with the *IEEE Transactions on Evolutionary Computation* as a flagship journal in this field, we owe to Walter and his faith in our community. He had no doubt that we would rise to the occasion. And now, sadly, the occasion is here and Walter is not, but I expect that Walter's influence on our community will be long lasting, and I hope that his ability to bring out more in individuals than they themselves might think possible will serve to encourage us to rise to future occasions all the more.

Acknowledgments

The author thanks George Bekey for inviting this paper. He also thanks Elsevier Science Publishers and the IEEE for permission to reprint portions of his prior publications as well as Eugene C. Wasson, Mike Boughton, Bill Porto, and Ankit Jain for their collaborative efforts. The collaborations with Wasson and Boughton were supported under Department of the Army, Grant No. DAMD17-96-1-6312. The content of the information presented does not necessarily reflect the position or the policy of the government, and no official endorsement should be inferred.

References

[1] R. Lippmann. "An Introduction to Computing with Neural Nets". *IEEE ASSP Magazine*, 4-22 April, 1987.

[2] K. Hornik, M. Stinchcombe and H. White. "Multilayer Feedforward Networks are Universal Approximators". *Neural Networks*, 2:359-366, 1989.

[3] T. Poggio and F. Girosi. "Networks for Approximation and Learning". *Proc. of the IEEE*, 78:9, 1481-1497, 1990.

[4] D.B. Fogel. *Evolutionary Computation: Toward a New Philosophy of Machine Intelligence.* 2nd ed., IEEE Press, Piscataway, NJ, 1999.

[5] D.B. Fogel and P.K. Simpson. "Experiments with Evolving Fuzzy Clusters". *Proc. of 2nd Ann. Conf. on Evolutionary Programming*, D.B. Fogel and W. Atmar (eds.), Evolutionary Programming Society, La Jolla, CA, 90-97, 1993.

[6] D. B. Fogel. *System Identification through Simulated Evolution.* Ginn Press, Needham, MA, 1991.

[7] D.B. Fogel, L.J. Fogel and V.W. Porto. "Evolving Neural Networks". *Biological Cybernetics*, 63:6, 487-493, 1990.

[8] V.W. Porto, D.B. Fogel and L.J. Fogel. "Experiments with Alternative Neural Network Training Methods: Back Propagation, Simulated Annealing and Evolutionary Programming". *IEEE Expert,* 10(3): 16-22, 1995.

[9] D.B. Fogel. "An Information Criterion for Optimal Neural Network Selection". *IEEE Trans. on Neural Networks*, 2:5, 490-497, 1991.

[10] http://breastcancer.about.com (see subpages as well).

[11] C.C. Boring, T.S. Squires and T.Tong. "Cancer statistics". *CA: Cancer J. Clin.* 43:7-26, 1993.

[12] P. Strax. *Make Sure You Do Not Have Breast Cancer.* St. Martin's, NY, 1989.

[13] Y.Z. Wu, M.L. Giger, K. Doi, C.J. Vybomy, R.A. Schmidt and C.E. Metz. "Artificial Neural Networks in Mammography: Application to Decision Making in the Diagnosis of Breast Cancer". *Radiology,* 187:81-87, 1993.

[14] L. Tabar and P.B. Dean. *Teaching Atlas of Mammography.* 2nd ed., Thieme-Stratton, NY, 1985.

[15] C.E. Floyd, J.Y. Lo, A.J. Yun, D.C. Sullivan and P.J. Kornguth. "Prediction of Breast Cancer Malignancy Using an Artificial Neural Network". *Cancer,* 74:2944-2998, 1994.

[16] P. Wilding, M.A. Morgan, A.E.Grygotis, MA. Shoffner and E.F. Rosato. "Application of Backpropagation Neural Networks to Diagnosis of Breast and Ovarian Cancer". *Cancer Lett.,* 77:145-153, 1994.

[17] D.B. Fogel, E.C. Wasson, E.M. Boughton and V.W. Porto. "A Step Toward Computer-assisted Mammography Using Evolutionary Programming and Neural Networks". *Cancer Lett.,* 119:93-97, 1997.

[18] J.A. Baker, P.J. Kornguth, J.Y. Lo, M.E. Williford and C.E. Floyd. "Breast Cancer: Prediction with Artificial Neural Networks Based on BI-RADS Standardized Lexicon". *Radiology,* 196:817-822, 1995.

[19] D.B. Fogel, E.C. Wasson, E.M. Boughton and V.W. Porto. "Evolving Artificial Neural Networks for Screening Features from Mammograms". *Artificial Intelligence in Medicine,* 14:317-326, 1998.

[20] D.B. Fogel, E.C. Wasson and E.M. Boughton. "Evolving Neural Networks for Detecting Breast Cancer". *Cancer Lett.,* 96:49-53, 1995.

[21] O.L. Mangasarian and W.H. Wolberg. "Cancer diagnosis via linear programming. *SIAM News,* 23:1-18, 1990.

[22] K.P. Bennett and O.L. Mangasarian. "Robust Linear Programming Discrimination of Two Linearly Inseparable Sets". *Optimization Methods Software,* 1: 22-34, 1992.

[23] W.H. Wolberg and O.L. Mangasarian. "Multisurface Method of Pattern Separation for Medical Diagnosis Applied To Breast Cytology". *PNAS,* 87:9193-9196, 1990.

[24] J. Zhang. "Selecting Typical Instances in Instance-based Learning". *Proc. 9th Int. Machine Learning Conf,* D. Sleeman and P. Edwards, eds., Morgan Kaufmann, San Mateo, CA, 470-479, 1992.

[25] J. Holland. *Adaptation in Natural and Artificial Systems.* Univ. Mich. Press, Ann Arbor, 1975.

[26] D.B. Fogel and A. Ghozeil. "A Note on Representations and Variation Operators". *IEEE Trans. Evolutionary Computation,* 1(2):159-161, 1997.

[27] D.H. Wolpert and W.G. Macready. "No Free Lunch Theorems for Optimization". *IEEE Trans. Evolutionary Computation,* 1(1): 67-82, 1997.

[28] G. Rudolph. "Reflections on Bandit Problems and Selection Methods in Uncertain Environments". *Proc. 7th Intern. Conf. on Genetic Algorithms,* T. Baeck (ed.), Morgan Kaufmann, San Francisco, CA, 166-173, 1997.

[29] W.G. Macready and D.H. Wolpert. "Bandit Problems and the Exploration/Exploitation Tradeoff". *IEEE Trans. Evolutionary Computation,* 2:1, 2-22, 1998.

[30] L. Altenberg. "The Schema Theorem and Price's Theorem". *Foundations of Genetic Algorithms 3,* L.D. Whitley and M.D. Vose (eds.), Morgan Kaufmann, San Mateo, 23-49, 1995.

[31] D.B. Fogel. "Phenotypes, Genotypes, and Operators". *Proc. 1995 IEEE International Conf. on Evolutionary Computation,* Perth, Australia, IEEE, 193-198, 1995.

[32] K. Chellapilla and D.B. Fogel. "Fitness Distributions in Evolutionary Computation: Analysis of Noisy Functions". *Applications and Science of Computational Intelligence II,* K.L. Priddy, P.E. Keller, D.B. Fogel, and J.C. Bezdek (chairs), SPIE, Bellingham, WA, 313-323, 1999.

[33] K. Chellapilla and D.B. Fogel. "Fitness Distributions in Evolutionary Computation: Motivation and Examples in the Continuous Domain". *BioSystems*, 54(1-2), 15-29, 1999.

[34] P. Nordin and W. Banzhaf. "Complexity Compression and Evolution". *Proc. 6th Intern. Conf. Genetic Algorithms*, L. Eshelman (ed.), Morgan Kaufmann, San Mateo, CA, 310-317, 1995.

[35] A. Jain and D.B. Fogel. "Case Studies in Applying Fitness Distributions in Evolutionary Algorithms. I. Simple Neural Networks and Gaussian Mutation". *Applications and Science of Computational Intelligence III*, D.B. Fogel, K.L. Priddy, and P.E. SPIE, Bellingham, WA, 2000.

[36] D.E. Rumelhart, G.E. Hinton and J.L. McClelland. "A General Framework for Parallel Distributed Processing". *Parallel Distributed Processing: Explorations in the Microstructure of Cognition*, D.E. Rumelhart and J.L. McClelland (eds.), MIT Press, Cambridge, MA, 1: 45-76, 1986.

[37] P.J. Angeline, G.M. Saunders J.B. Pollack. "An Evolutionary Algorithm that Constructs Neural Networks". *IEEE Trans. Neural Networks*, 5(1): 54-65, 1994.

[38] X. Yao. "Evolving Neural Networks". *Proc. IEEE*, 87:9, 1423-1447, 1999.

17

Road Map and Issues in Collaborative Visualization

Alex Pang and Craig M. Wittenbrink
University of California, Santa Cruz
NVIDIA, Santa Clara, CA 95064
pang@cse.ucsc.edu

Summary. A majority of the collaborative systems today are still in 2D. This paper highlights several issues and challenges in extending support to 3D collaborative workspaces. In particular, we focus on collaborative visualization systems that support shared 3D virtual workspaces. Several features unique to 3D collaborative visualization systems can be identified. For example, the number of participants in a session is usually relatively small; the data sets and the amount of data transferred in a session are typically large; and because each participant may be located anywhere within the 3D virtual workspace, each participant may have his own viewpoint or perspective. These properties help constrain and shape the design of 3D collaborative visualization systems. This paper presents two related 3D collaborative visualization systems called CSpray and PET SLUG, and discusses the particular issues and challenges shared by other 3D collaborative visualization systems. We provide a road map from the insights developed from our research, to the development of 3D collaborative visualization.

1. Introduction

At present, most scientists share data and research results through conference and journal publications, a process that takes weeks or months. Slowly, the tools that are used for data and information sharing have progressed from paper mail, fax, email, web browsers, to specialized collaboration tools. In addition, recent developments in teleconferencing allow geographically separated colleagues to discuss ideas using voice and video in real time.

There are many systems designed to support general multimedia interaction among users particularly from the CSCW (computer supported collaborative work) community. One notable work that allows collaboration of stand-alone X11 applications is CECED [6]. This was later extended to support synchronous collaboration over the WWW in the COMET system [9]. A much smaller body of work focuses on collaborative systems that support 3D visualization. In general, these works are research systems, prototypes, or initial market offerings in Mechanical CAD (Computer Aided Design). Among these are: Shastra [1] which supports multiple collaborative applications; Studierstube [10] includes augmented reality in both fixed and mobile collaboration in a number of application domains; COVISE [21] which replicates data from remote participants and performs local rendering; the VisAD [12] Java component library was used in in a collaborative geovirtual environment [14] where a Java application, called Talk- Server, listens and handles collaboration events from a predetermined port on a server; similarly the Habanero system was used to convert a stand-alone GIS application into a collaborative one [11]; Wood et al. [23] describe how the data flow approach to visualization can be extended to a collaborative setting. Industrial CAD tools vendors have fielded collaborative systems including OneSpace (CoCreate), Pivotal (Centric Software), e-Vis.com (EAI), and Concept Works (Reality Wave) [16]. The Electronic Visualization Laboratory (www.evl.uic.edu) is another place where state of the art visualization is happening. One of their hallmarks is work on networked CAVE environments [13]. Some recent advances in collaborative augmented reality are reported in [3], while a good source for additional information on 3D networked graphics can be found in *www.siggraph.org/s2001/conference/courses/crs26.html.*

The works listed above are still the exception. Most visualization systems operate in single user mode. With single-user tools, visualizations are created from one graphics workstation. Users may run remote modules, for example on a supercomputer, but they do not interact with other users in the creation of the visualization. In contrast, extending single-user visualization tools into *interactive collaborative scientific visualization* settings allow multiple investigators to share data, views, manipulation sequences and to participate in the creation of the visualization products across the network.

Designing a system for collaborative scientific visualization requires a new look at what it means to have groups concurrently visualize data. There are many new design issues, including the user interface, authorization levels on data and users' control. This paper reports the evolution of our efforts in developing a 3D collaborative visualization system. Our new solutions for a scientific collaboration environment provide:

- Synchronized workspaces for group generated visualizations.

- Private workspaces for visualizations generated independently of a collaboration group.
- Coordination of interactive media, such as video, audio, and text
- Local data integrity, protection, and independence from the collaboration.

We organize the rest of the paper in the following manner. Section 2 discusses parameters for general collaborative problem solving environments; Section 3 identifies the relevant subset for supporting 3D collaborative visualization environments and their particular constraints. We then discuss the evolution of our development effort for providing a 3D collaborative visualization environment in Section 4. Finally, we conclude with how this work is important to developers of collaborative problem solving environments.

2. Parameters for Collaborative Systems

Macedonia and Zyda [15] describe a taxonomy for networked virtual environments. Their taxonomy is centered around network communication (bandwidth, distribution, latency, and reliability), views (synchronous vs. asynchronous), data (replicated, shared, centralized, distributed), and processes (local, remote). We describe several more categories that can be used to distinguish collaborative systems. The parameters affect the requirements for the application interface(s), networking support, and networking communication. While further studies are needed to evaluate the impact of these parameters on the design of collaborative systems, the list below can be used as a road map in developing, analyzing, and comparing systems. If too many parameters are used to categorize collaborative systems the number of examples only sparsely populates the taxonomy. Therefore, some of the parameters in the table have been grouped together.

Table 1 shows some of the parameters and their range of values. For example, the first parameter is the number of participants in a collaborative setting that may range from two to thousands. The expected number of simultaneous users impacts the choice of network topology, and may depend on the type and interactivity of the application. Other parameters are the communication approach, where synchronous interactions imply that participants are present during a session which is also shorter. Asynchronous communication approaches tend to run for an extended period of time as all participants do not need to be simultaneously present. Participants may collaborate within the same room, across countries, or across continents.

The bandwidth and latency of the network are controlled by the participant locality as the network services change from LAN (Local Area Network) to WAN (Wide Area Network). Dealing with communication delays and different time zones are common design issues. Different participant locality allows for different modes of interaction. Examples include, when participants are in the same place, augmented reality [8, 3] and virtual workbenches may be used, and when participants are remote CAVE environments [7] may be used.

Parameter	Value
Number of participants	2 or more
Communications	synchronous/asynchronous
Participant locality	same room to different time zone
Spatial dimensionality	2D/3D
Streaming content	audio, video, text, graphics
Collaboration support	dedicated, application-independent
Operating system support	yes/no, WWW, java, ...
Network topology	star, tree, graph
Network protocol	UDP/TCP/IP
Network security	yes/no
Process execution	local/remote
Data sharing	yes/no, raw/representative
Data privacy	yes/no, permission
Application domain	Flow visualization, CAD, games, education, ...

Table 1. These parameters allow one to distinguish one collaborative system from another and can form the basis for a classification.

A collaborative setting may range from passing simple state information, transfer of derivative products, to full exchange of raw data, processes, and ideas. The spatial context for this exchange may be in 2D or 3D. Earlier collaborative applications were mostly textual in nature. Increasingly, the trend has been towards 3D workspaces and a richer set of multi-streaming content interactions.

Applications need a layer of software to enable them to communicate with each other and allow users to collaborate with each other. The collaboration support in some systems are intimately tied with the application (such as the virtual wind tunnel [5]), while others may be more independent of the application (such as Shastra [2]). Shastra supports multiple collaborative applications using the same infrastructure.

Table 1 also summarizes several network parameters. The network topology may be specified for the physical and logical network. Participants may be fully connected to each other using peer-to-peer or indirectly through client-server. Data among participants may be broadcast to everyone, multicast to interested parties, or communicated directly using unicast. The choice impacts the fault tolerance of the connections, amount of traffic flow, and the scalability of the collaborative system. Different types of streams may use different topologies, for example control could be broadcast, while data updates could be multicast. Data may be transferred using UDP or TCP/IP whether middleware or custom socket connections are used. The main concern is reliability, speed, and packet ordering, if necessary. In fact, these three could be parameters in their own right. One advantage is that they can be easily, discretely quantifiable.

Process execution may occur on the local host or a remote machine. This could range from all hosts being equals to a client-server or master-slave arrangement when an expensive resource (such as a supercomputer) is involved in the session.

Data may be replicated on the machine of each participant, maintained at a central site, or at the site of the participant that owns the data. Data may be shared directly or indirectly. That is, either raw data or its representative form (such as visualization primitives: polygons or rendered images) may be shared. The different levels of data sharing can help alleviate concerns about sensitive data sets and allow resource allocation tradeoffs such as data transmission and local processing. Privacy and security can also occur at different levels. Is there an authentication process before one can join a session? Once one has joined a session, are there additional permissions that need to be requested before one can access sensitive data? The absence of some of these features may prove unacceptable for certain applications or users.

Some aspects of collaboration systems are not neatly parameterized. Examples include collaboration aids, floor control, and session management. Collaboration aids facilitate the discussion among participants. There is a wide range of collaboration aids including: avatars, linked cursors, shared views, and side-sessions. Some of these aids are more general, while others are application and/or task specific. Floor control allows access to shared resources in a collaborative system. It can range from a free-for-all to very regulated floor control support. The main issues are fairness and unobtrusiveness of the floor control strategy. Aside from the floor control support provided, another distinguishing factor is whether floor control is provided by a middleware layer or directly by the application. The session management tasks include maintaining: a list of active sessions that participants can join in, a list of participants in each session, and the state of each collaborative session. Keeping track of the state will allow late-comers to be brought up to date, and early departures to exit nicely. The session management service may be provided from a central server, or it may be distributed. Aside from the session management services provided, another distinguishing factor is whether the services are provided by a middleware layer or directly by the application.

3. 3D Collaborative Visualization Systems

Since this paper's focus is on 3D collaborative visualization systems, we look at a number of design issues and constraints particular to these systems. Then, in the next section, we describe two systems that we have built and how they relate to the parameters described.

1. *Navigation and collaboration aids.* With the added dimension of space (beyond the normal 2D space in most CSCW systems), it is much easier to get lost within the shared virtual workspace. A number of 3D navigation and collaboration aid alternatives are possible such as earcons [4] 3D avatars and eyecons (3D icon shaped like an eyeball), 3D pointers, examination and fly through/around modes, and private or shared windows and views.

2. *Managing visual complexity and consistency in shared views.* Typical visualizations will include things like contour lines, pseudo-colored slices, 3D isosurfaces, and streamlines. A collaborative session can very quickly produce displays with too much clutter and inconsistent views. Some of the mitigation approaches include multiple windows, private versus public windows, and the ability to form sub-sessions.

3. *3D annotation tools.* A shared 2D pointer in a 3D virtual workspace is not sufficient because each participant can potentially have a different viewpoint. A 3D pointer provides the ability to point at a particular location in space without ambiguity for multiple viewers from different viewpoints. Coupling the 3D pointer with annotation capability allows participants to tag and label interesting features of their visualization.

4. *Bandwidth issues.* The amount of data exchange in a 3D scientific visualization session is significantly larger than typical CSCW sessions. On the other hand, there is more tolerance for lags and slight delays. Some tradeoffs can be made to address this factor through compression and different types of data exchange (such as raw data, geometry data, and images).
5. *Data ownership and access.* While there is nothing technically different about data ownership and access issues for 3D collaborative visualization systems compared to other CSCW systems, participants of a collaborative visualization session are found to be typically very protective of their raw data sets. For example, the data to be visualized may have been very costly to produce or gather, or the data may be confidential or still unpublished. Whatever the reasons, participants must be assured of controlled access to their data in order to encourage their use of collaborative systems. This concern is not different from those of providing web security to enhance e-commerce.
6. *Floor control.* Shared objects may be in the form of 3D objects. Floor control strategies, while attempting to be fair and intuitive, must also be sensitive to the fact that some of these objects may not be visible from the current viewpoint. Indications of who has the floor on a shared object should ideally be shown together with the rendering of the shared object rather than looking up some other pulldown menu.

4. CSPRAY and PET SLUG

We now describe two collaborative visualization systems that we have developed that address the concerns raised in the previous section. The first system is called CSpray [19] which was developed around 1996. The second system is called PET SLUG [20] and is a follow on to the CSpray system. Details are available in the references cited above as well as www.cse.ucsc.edu/research/avis/envis.html. In these descriptions, we discuss the evolution and changes made to PET SLUG based on lessons learned in developing CSpray. We have taken user feedback, technical challenges, and redesign opportunities to create a road map for the PET SLUG enhancements.

4.1 CSPRAY

CSpray is based on Spray Rendering [17] where virtual spray cans emit smart particles (*sparts*) to create a visualization. The spray paint metaphor provides intuitive control for a variety of visualization methods to interactively explore and analyze data. As the *sparts* enter the data space, they seek features and map them to visualization primitives. Figure 1 illustrates the effects of spraying a can that generated contour lines and another can that generated points colormapped to data values. *Sparts* can be simply programmed for different behavior by changing target features, visualization actions, position update rules, and termination rules. There is also a Mix&Match [18] graphical interface for interactively creating new types of *sparts*.

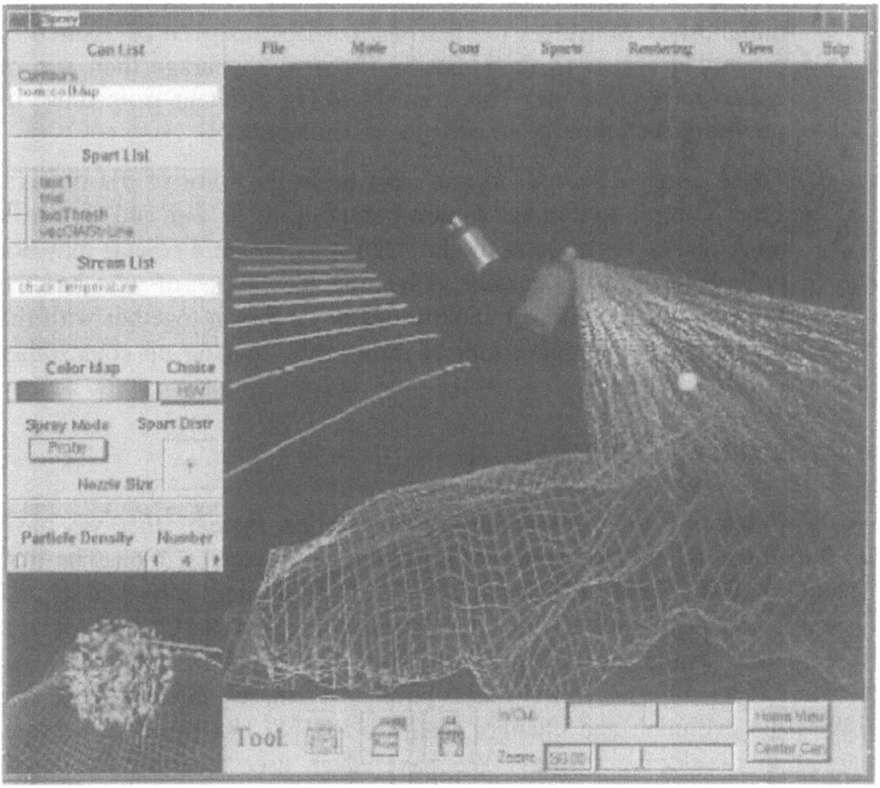

Figure 1. Spray Rendering workspace. The main window shows the user's current viewpoint, while the lower left window shows the viewpoint from the currently active spray can.

CSpray is Collaborative Spray rendering [19]. It allows geographically distributed users to interactively create and analyze visualizations within a shared virtual 3D workspace. Data being visualized do not have to be replicated at every site nor centralized in one site. Instead data are distributed, owned, and managed by the local hosts. Participants in a collaborative session can secure their raw data, and yet allow collaborators to access data indirectly through visualizations and images. Clutter in the shared displays is reduced by distinguishing between private and public visualization objects and views. CSpray also has its own floor control manager for handling access to shared objects (spray cans) as well as a simple session manager that brings late comers up-to-date and handles early terminations. Finally, it supports recording and playback of session activities.

The CSpray architecture uses peer-to-peer communication with streams. While this limits the scalability, visual clutter is often the practical limiting factor. There is no central server in a CSpray session; a new participant can join the session by connecting to any existing participant in the session. Participants may drop out or rejoin as long as there is one session member. To support this, each participant maintains their own list of active participants and their resources such as position, control of shared objects, and visualization primitives. When a participant leaves a session, that event is broadcast to its list of active participants. Those members in turn will update their list of active participants. When a participant joins a session by connecting to a specific member, that member broadcasts the information about the new participant to its list of active members, and also brings the new participant up to date with a list of current members, their positions, control over shared objects, and visualization primitives. In other words, the collaboration code is handled within CSpray itself. Separate streams are provided for data and control with tags identifying whether events are local or remote.

4.1.1 Collaboration Aids

Several aids are provided to facilitate group collaboration within a virtual 3D environment.
These are:

1. *Representation of session participants.* Each active participant is represented by an *eyecon*, a 3D icon shaped like an eyeball, for position and view direction, as well as in a pulldown Views menu.
2. *Sharing views.* CSpray provides shared views (Figure 2) with a *public* window, and private views with a *private* window. Shared

windows are controlled by selecting another participant's viewpoint or following a briefing driven by one member. Strictly speaking, CSpray provides a *WYSIWILYS* paradigm – What You See Is What I Let You See. Individual members have control over parts of the data and visualization they deem are necessary or ready for sharing. That is, the creator of a spray can (and hence the owner of the data set associated with the spray can) has a choice of making the spray can public or private at any time. Only the public cans and their corresponding visualizations are shown in the public windows of other users. This also helps to reduce clutter in the shared virtual workspace.

3. *Sharing data.* Data sharing is explicitly supported with three different levels: raw data, visualization primitives, and visualization images. These levels of sharing allow protecting sensitive raw data, and matching client's rendering capabilities. Primitives can be sent instead of raw data, and slow clients can use visualization images from others.
4. *3D pointers.* Traditional cursors are inadequate for collaboration with multiple windows and users, so we provide 3D pointers attached to spray cans and controlled in a similar fashion. Users can annotate pointed out regions of interest with text that it is readable from any direction.

4.1.2 Floor Control

In CSpray, public spray cans are the only shared objects. Each public can has a floor manager and a transferable floor controller. Floor management is locally *static* because the ownership of the public can always resides on the same machine where it was created. On the other hand, floor control is *distributed* and does not have to be requested from one central source thereby improving the scalability of the number of shared objects. CSpray uses a traffic light analogy for its floor control strategy: red (*taken*), yellow (*requested*), and green (*owned*). When someone requests an owned can, the can label turns from red to yellow to green, while the owner sees the can label turn from green to yellow to red. Control may be explicitly released or it times out.

4.2 PET SLUG

PET SLUG [20] is the follow-on to the CSpray collaborative visualization system. PET is an acronym for Products, Elements, and Tools which reflect the architecture of the new system. Application users (such as meteorologists) typically think in terms of visualization products (such as weather forecasts), while visualization users think in terms of what visualization tools (including contour plotting and streamline tools) are needed to generate the product. Visualization developers, on the other hand, think of components or elements that make up a tool. SLUG is for the banana slug mascot of UCSC, and hence the name of our new system. PET SLUG reflects a number of changes based on our experiences with CSpray.

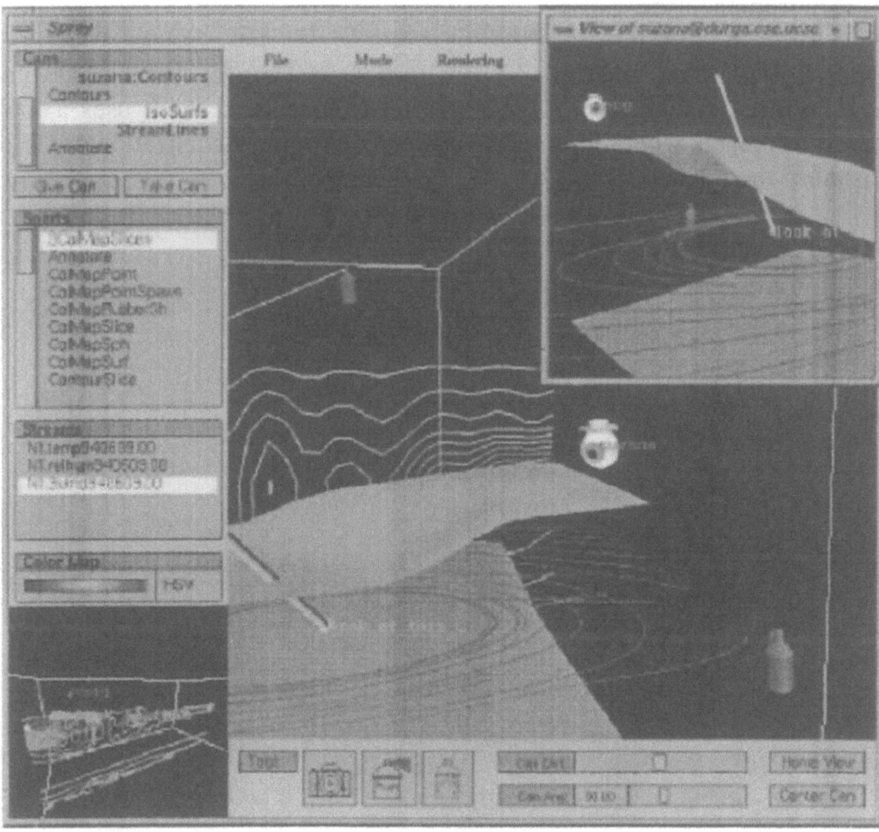

Figure 2. CSpray workspace in a two person collaboration. The main window is Pang's view where Suzana's eyecon is visible. The public window on the upper right contains Suzana's view where she can see Pang's eyecon. A 3D pointer with textual annotation is visible together with isosurfaces, contours, and streamlines.

The new system is being developed using OpenGL, C++, and xforms [24] for improved portability. In contrast, Cspray used IrisGL, C, and forms. PET SLUG uses dynamic shared object libraries to facilitate phased incremental release of the system. It is also a tool-based system as opposed to a mode-based system allowing users to activate different visualization tools as needed. While most of the changes in PET SLUG are in the visualization and interaction code, there are also significant modifications for collaboration. It now supports sub-sessions where two or more participants can temporarily break off from the main collaborative session for a side discussion and then come back. Floor control and session management code are also better separated from the visualization code. Finally, in preliminary tests, the system is perceived as being easier to use while the extensibility is hidden from the user.

PET SLUG is an interactive visualization system for space and time synchronized visualization of realtime environmental data. It uses a tool-based approach where users activate one or more visualization tools to visualize their data.

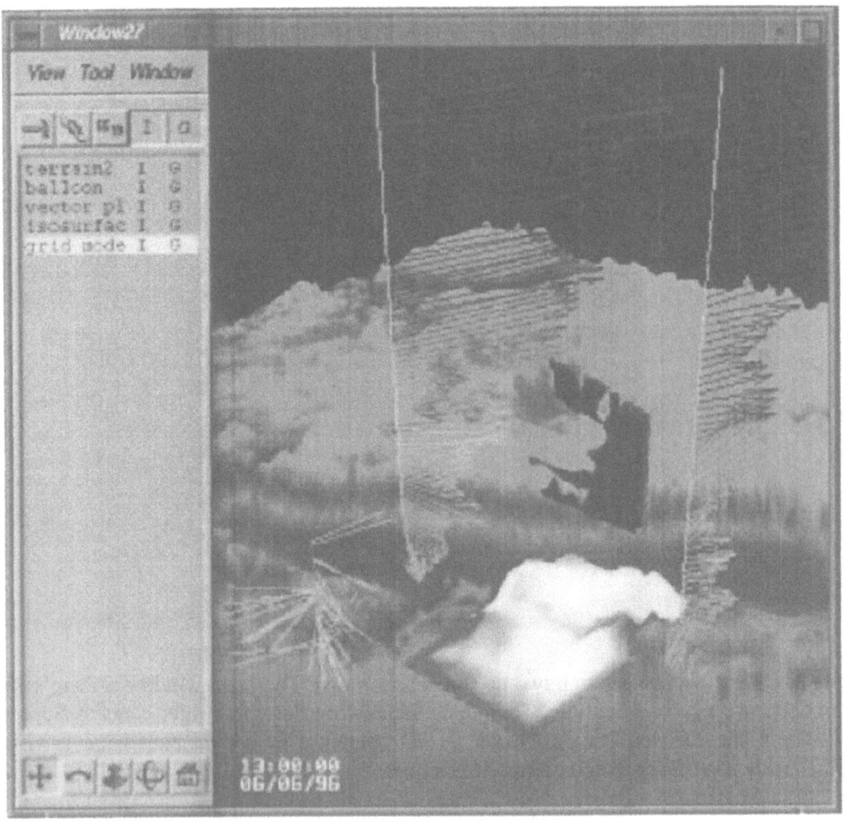

Figure 3. Graphical interface of PET SLUG showing visualizations of a variety of realtime environmental data.

Figure 3 shows the graphical interface of PET SLUG. Among the major components of PET SLUG are: distributed database query interface, geometry library, visualization tools, and multiple graphics windows for collaboration.

4.2.1 Distributed Database Query Interface

A graphical user interface is provided to the user to query the database via an application programmer's interface [22] that isolates the visualization application from the database development efforts. Queries are specified by three general characteristics: region of interest, time period of interest, and environmental parameter (such as pressure and humidity). The first two are specified via the *region selector* (see Figure 4). Different map projections are also supported here. Local data sets may also be used in the collaboration session with support of HDF files.

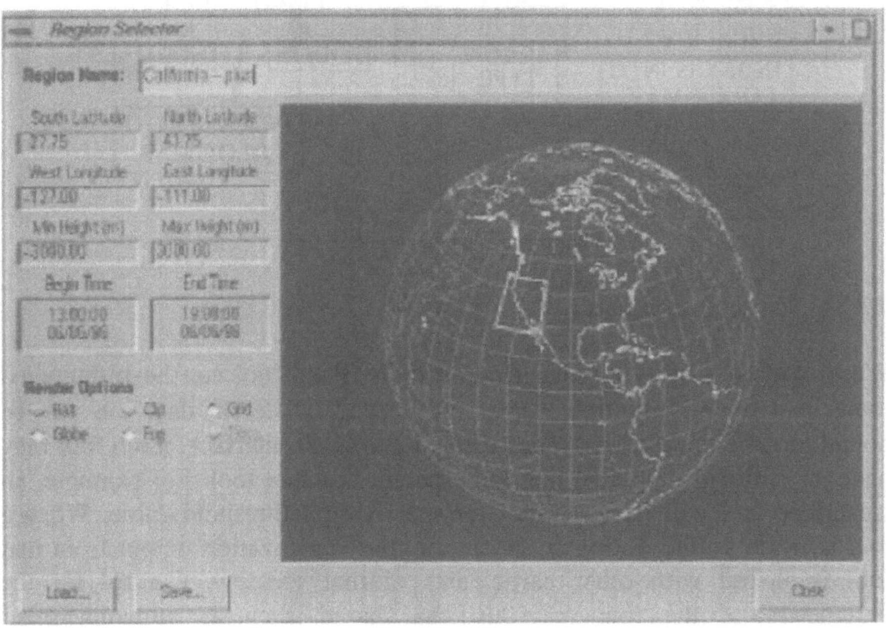

Figure 4. Region selector for specifying the region and time period of interest.

4.2.2 Geometry Library

The geometry library operates behind the scenes to provide support for tools requesting different types of geometry to be mapped to data values and sent to the renderer. Among the items supported in the library are: 3D text, color tables, various types of glyphs, polygon base class, lines, and points. These items are used by individual tools to create a hierarchical display list that contains geometry nodes to represent a particular visualization. Figure 5 shows the geometry tree for a tool displaying arrows and a polygonal surface, both using a common colormap. The display list is then sent to the renderer for display in a specified window. During a collaborative session, the display lists are also passed to other participants and maintained for consistency.

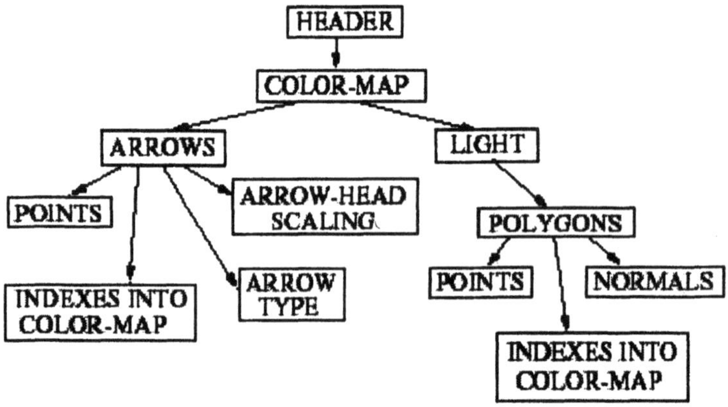

Figure 5. A sample geometry tree

4.2.3 Visualization Tools

We provide a variety of visualization tools. Each tool can be instantiated more than once if multiple copies are needed. Different data sets can be bound to different tool instances using a graphical interface. Each tool may have an optional graphical interface specific for that tool. For example, an isosurface tool will have a slider for specifying a threshold value. When a tool is made public during collaboration, the visualization output from that tool is shared with other participants of that window. Control for the parameters of the tool may also be shared such as allowing other participants to change a threshold value in order to generate a new isosurface for the data. Visualization tools include: contours, interpolation for spatial and temporal data, isosurfacing, image display, terrain, and vector plotting.

4.2.4 Collaboration and Multiple Graphics Windows

During a visualization, the user may create one or more windows for the renderers to send their output. Each window may have its own region, camera, and associated set of tools. Additionally, these objects may be shared across multiple windows. The graphics in a window, with its set of tools and datasets, constitute what we refer to as a visualization product. Thus, generating several graphics windows corresponds to generating multiple visualization products.

A list of active windows is visible to the user. For each of these windows (example in Figure 3), a list of tools can also be seen. Aside from this (not show in Figure 3), a list of participants in that window can also be seen. That is, collaboration is maintained for each window. In a collaborative session, different windows (and their associated tools and widgets) can be made public. A participant may participate in more than one collaborative session – perhaps one session with colleagues A, B, and C in one window, and another session with colleagues A and D in another window. In fact, this is also how we support sub-sessions where a small splinter session may form and dissolve from a main collaborative session. Like CSpray, participants can form a collaboration by joining any participant (non-central server). Unlike CSpray, participants need an extra step and specify which one(s) of the public windows (sessions) they want to join. In effect, each user maintains the most recent list of participants. Windows that are available for collaboration are advertised in such a list and different participants may join and leave as desired. When a new participant joins, their window is brought up to date with the latest set of tools in that window, together with all the graphics in it. When a participant leaves, their information as well as the tools they contributed to that window are removed from the remaining participants.

Currently, there is no direct support for merging windows or sessions. Remember that aside from being a unit for collaboration, windows are also used to hold the visualization product. Hence, conceptually merging two potentially different products may not make sense. However, in the context of merging a sub-session with a larger session, one can achieve similar results by one of two ways: (a) replicate the tools in the sub-session window into the main session window, or (b) ask the participants of the main session to join the sub-session window – the sub-session is already public anyway.

4.3 Summary

Having described the basic features of both CSpray and PET SLUG, we now look at how they relate to the parameters described in the earlier sections. Table 2 provides the parameter values for CSpray and PET SLUG. Unless noted specifically, the classification applies to both systems.

The tools are flexible, support a moderate number of users in a 3D collaboration, and maintain a peer-to-peer network connection and session management approach. CSpray can record all the events in a session, which can be played back at a later time. During playback, new users can "interact" with that session. This can be viewed as a form of asynchronous collaboration. We use this primarily for training purposes. Data is distributed and privacy is enforced through permissions to access data. Spray cans and tools can be shared. However, only the control of PET SLUG tools can be shared. As an example, remote collaborators can change the threshold of an isosurface tool.

Parameter	Value
Number of participants	2 to 5, limited by visual cluster
Communication	synchronous
Participant locality	same room to different time zone
Spatial Dimensionality	3D
Streaming content	3D billboard text, graphics
Collaboration support	dedicated
Operating system support	unix-based platform
Network topology	peer-to-peer
Network protocol	UDP (data), TCP/IP (events)
Network security	none specific to application
Process execution	local to each host
Data sharing	representative
Data privacy	permission
Application domain	scientific visualization

Table 2. Parameters for CSpray and PET SLUG.

5. Conclusions

We have enumerated and described several key considerations in extending collaborative problem solving environments to support distributed 3D visualization. These are then demonstrated in two systems: CSpray and PET SLUG. Within these collaborative visualization systems, users interactively create visualizations from distributed data sets. A number of collaboration aids facilitate navigation and interaction within a shared 3D virtual workspace. Our work to date uses shared views using multiple windows, and floor control for tools that affect content of those windows. We have also developed protocols for peer-to-peer communication, and shown their effectiveness in interactive collaborations. Protocol developments included the use of mixed streams of geometric primitives, compressed geometric primitives, and image compression to help improve performance and handle disparate client performance. Many of the issues we addressed in the development of our systems have created solutions that have direct parallels in commercial collaborative software such as [16]. As software vendors continue to add more collaboration features, they will come upon 3D collaboration issues we have addressed. Our CSpray and PET SLUG design decisions therefore provide a road map for enhancing 3D visualization and collaboration tools.

Acknowledgements

We would like to thank the members of the Advanced Visualization and Interactive Systems laboratory at UC, Santa Cruz for their feedback and suggestions. This work is funded in part by DARPA grant N66001-97-8900, NASA Cooperative Agreement NCC2-5281. LLNL Agreement No. B347879 under DOE Contract No. W-7405-ENG-48, and ONR grant N00014-00-1-0764.

References

[1] V. Anupam and C. Bajaj. “Collaborative Multimedia Scientific Design in SHASTRA”. *Proc. of the ACM Conference on Multimedia Systems*, 447–480, August 1993.

[2] V. Anupam and C. Bajaj. “Shastra: Multimedia Collaborative Design Environment”. *IEEE Multimedia*, 1(2):39–49, Summer 1994.

[3] M. Billinghurst and H. Kato. “Collaborative Augmented Reality”. *Communications of the ACM*, 45(7):64–70, July 2002.

[4] M. Blattner, D. Sumikawa, and R. M. Greenberg. "Earcons and Icons: Their Structure and Common Design Principles". *Human-Computer Interaction*, 4(1), 1989.

[5] S. Bryson and C. Levit. "The Virtual Wind Tunnel". *IEEE CG&A*, 12(4):25–34, July 1992.

[6] E. Craighill, R. Lang, M. Fong, and K. Skinner. "CECED: A System for Informal Multimedia Collaboration". *Proc. of the ACM 1993 Multimedia Conference,* Anaheim, CA, August 1993.

[7] C. Cruz-Neira, D.J. Sandin, and T.A. DeFanti. "Surround-screen Projection-based Virtual Reality: The Design and Implementation of the CAVE". In James T. Kajiya, editor, *Computer Graphics (SIGGRAPH '93 Proc.)*, 27: 135– 142, August 1993.

[8] S. Feiner, B. MacIntyre, M. Haupt, and E. Solomon. "Windows on the World: 2D Windows for 3D Augmented Reality". *Proc. of the ACM Symposium on User Interface Software and Technology*, 145–155, ACM Press, November 1993.

[9] T.J. Frivold, R.E. Lang, and M.W. Fong. "Extending WWW for Synchronous Collaboration". *Computer Networks and ISDN Systems*, 28(1-2):69–75, Dec 1995.

[10] A. Fuhrmann, H. Loffelmann, D. Schmalstieg, and M. Gervautz. "Collaborative Visualization in Augmented Reality". *IEEE Computer Graphics and Applications*, 18(4):54–59, July-August 1998.
URL: www.studierstube.org/research master.php.

[11] E. Grossman and A. Woss. "Integrating a GIS Application into the Habanero Collaborative Framework". In R. Laurini, editor, *TeleGeo99, First International Workshop on Telegeoprocessing*, 164–167, 6-7 May 1999.

[12] W. Hibbard, C.R. Dyer, and B. Paul. "Display of Scientific Data Structures for Algorithm Visualization". In A.E. Kaufman and G.M Nielson, editors, *Proc. IEEE Visualization '92*, IEEE Computer Society Press, 139–146, 1992.
URL: www.ssec.wisc.edu/ billh/visad.html.

[13] J. Leigh, A.E. Johnson, T.A. DeFanti, et al. "Review of Tele-immersive Applications in the CAVE Research Network". In L Rosenblum, P. Astheimer, and D. Teichmann, editors, *Proc. IEEE Virtual Reality*, IEEE Computer Society, 180–187, March 1999.

[14] A. M. MacEachren, I. Brewer, and E. Steiner. "Geovisualization to Mediate Collaborative Work: Tools to Support Different-place Knowledge Construction and Decision-making".
URL: hero.geog.psu.edu/products/ICC16009amm.pdf.
URL: www.geovista.psu.edu/research/collaborativevisualization.

[15] M.R. Macedonia and M.J Zyda. "A Taxonomy for Networked Virtual Environments". *IEEE Multimedia*, 4(1):48–56, Jan-Mar 1997.

[16] J. Mackrell. "Collaboration in Product Development". *Desktop Engineering (DE Online)*, 5(4), Feb 2000.
URL: www.deskeng.com/articles/00/Feb/index.htm.

[17] A. Pang. "Spray Rendering". *IEEE Computer Graphics and Applications*, 14(5):57–63, 1994.

[18] A. Pang and N. Alper. "Mix & Match: A Construction Kit for Visualization". In *Proc.: Visualization '94*, 302–309, IEEE Computer Society, 1994.

[19] A. Pang and C. M. Wittenbrink. "Collaborative 3D Visualization with CSpray". *IEEE Computer Graphics and Applications*, 32–41, March 1997. (Special issue on: 3D and Multimedia on the Information Superhighway.)

[20] E. Saxon, Z. Wood, M. O'Neil, C. Oates, J. Story, S. Djurcilov, and A. Pang. "Integrated Visualization of Realtime Environmental Data". In *Proc. of the Spring Conference on Computer Graphics*, 135–143, Comenius University, Bratislava, 1997.

[21] A. Wierse. "Performance of the COVISE Visualization System Under Different Conditions". In *Proc. of the SPIE*, Visual Data Exploration and Analysis II, 2410: 218–229, 8-10 February 1995.
URL: www.hlrs.de/organization/vis/covise.

[22] C. M. Wittenbrink, E. Rosen, A. Pang, S. Lodha, and P. Mantey. "Realtime Database Support for Environmental Visualization". In G. Grinstein, U. Lang, and A. Wierse, editors, Lecture Notes in Computer Science, Second Workshop on Database Issues for Data Visualization, Springer Verlag, 1183: 111–130, 1996.

[23] J.D. Wood, H. Wright, and K.W. Brodlie. "Collaborative Visualization". In R. Yagel and H. Hagen, editors, *Proc. IEEE Visualization '97*, IEEE Computer Society Press, 253–260, 1997.
URL: www.comp.leeds.ac.uk/vis/jason/vis97/vis97.html.

[24] T.C. Zhao. XFORMS home page.
URL: http://bragg.phys.uwm.edu/xforms.

18

Intelligent Interfaces for Mission-critical Systems

Hamid S. Kohen
Research Associate
California Institute of Technology, Pasadena, CA 91125
kohen@jpl.nasa.gov

Justin A. MacDonald
Research Assistant
Purdue Universit, West Lafayette, IN 47907
macdonja@purdue.edu

Christian Yoedhana
Research Assistant
University Of California at Los Angeles, CA 90095
jdb@psych.purdue.edu

J. D. Balakrishnan
Associate Professor of Psychological Sciences
Purdue University, West Lafayette, IN 47907
jdb@psych.purdue.edu

Summary. Although expert systems continue to develop at a rapid pace, human experts still make most of the important decisions in our society. When these experts make a mistake, society is often quick to cry foul. Experimental work in the behavioral sciences, however, has shown that humans often commit errors for the same reasons that make them superior to artificial systems to begin with, i.e., because they have acquired a wealth of problem-solving heuristics and intuitions that work well in general but are rarely optimal in a specific problem context. In this chapter, we describe some recent work on the development of intelligent interfaces that can anticipate the circumstances of human errors arising from naturally formed heuristics, or cognitive illusions, and provide appropriate feedback and training to eliminate them. The approach involves identifying the conditional probabilities relating objective events to subjective impressions of event probabilities and determining whether the decision making strategy implied by these conditional probabilities is suboptimal.

1. Introduction

New technologies in the engineering and computer sciences are often beacons of opportunity for human-computer interface (HCI) designers. Personal computers, graphical user interfaces, the computer mouse, head and eye tracking devices and virtual reality have all found their place in HCI research laboratories and our everyday lives. And of course, some researchers found these more quickly than others.

Walter Karplus was one of the faster ones on the draw. His work for the NASA Dryden Flight Research Center and later for the Jet Propulsion Laboratory (JPL) was in many ways an aggressive attempt to bring together various kinds of frontier technologies, such as simulated 3-D sound, automated speech recognition, computer imaging technologies and virtual reality, to create a better HCI model for situations in which the model is crucial, i.e., in the air traffic and flight test control rooms at the NASA centers.

In these high risk applications, human errors and their consequences can be as drastic as they are difficult to predict. When they do occur, the unfortunate perpetrator – and sometimes an entire program – is often stigmatized and deemed unfit for the job. However, many of these errors can ultimately be traced not to mere negligence or lack of proper training on the part of an individual or program but instead to fundamental aspects of normal human cognition and reasoning capabilities. Recognizing this crucial aspect of the problem, Prof. Karplus developed a keen interest in the findings of psychologists relating to the limits of cognition and the illusions of reality that apparently all of us are susceptible to. In this chapter, we review some of the work on intelligent, "cognition-based" interfaces that Prof. Karplus initiated, guided and continued to support until he lost his life to cancer.

1.1 Intelligent Systems and the Human Computing Center (HCC)

The human-centered approach to interface design is by its nature interdisciplinary, integrating as it must the computer hardware and software developed by teams of experts and designed with an understanding (or at least, an attempt to develop one) of the key differences between the cognitive and computational abilities of human and machine. The ultimate objective of such an endeavor is a principled design scheme that, in effect, incorporates a model of human attention, motivation, memory, knowledge and concept structures, decision-making styles, principles of learning, and sensory perception. In short, a system that takes into account all of the known regularities of human experience and behavior. Its success can be gauged by the robustness, speed, and reliability of the entire system, including the human operator.

One innovative feature of the system we are developing at JPL's Human Computing Center (HCC) is the introduction of different forms of intelligent feedback that provide the human user information about potential biases that may affect performance in the context of a given problem. Some of this feedback cannot be hardwired or preprogrammed – the system must learn about idiosyncratic properties of the user and adapt its outputs to them. The software therefore incorporates learning algorithms (neural network-based) that attempt to discover the patterns that best represent the categories that a human operator uses to represent his or her objectives, properties of the environment and the problem context (e.g., the movement patterns of aircraft in an air traffic display) and, most importantly, relationships between the user's information states and their contingent probabilities for error.

1.2 Cognitive Illusions and Their Relevance for NASA

Cognitive psychologists have been aware for some time that expert and novice decision makers are subject to some strikingly consistent and, therefore, predictable errors in judgment. These errors, which often lead to seemingly irrational and ultimately incorrect decisions, are not due to ignorance, lack of training, low intelligence, poor motivation or fatigue – they are mistakes that arise due to certain imperfections in the fundamental mental mechanisms of normal human thought processes. In an important sense, they are analogous to the more familiar optical illusions of visual perception, in that they are due to an apparently irresistible tendency to apply general heuristics to tough problems, sometimes even when a specialized, logical analysis would produce a more veridical result.

Some of the pioneering work relating to cognitive illusions was carried out in the 1960s and 1970s by Amos Tversky and Daniel Kahneman and has been elaborated by a host of research psychologists since then [1-5]. Piattelli-Palmarini [5] provides a nice introduction to the subject (and, perhaps, the hook that first caught Prof. Karplus' attention). According to Edwards and von Winterfeld [see 1], every cognitive illusion contains the following elements:

- A formal rule that specifies how to determine a correct answer to a question
- The question normally includes all information required as input to the formal rule
- A judgment, made without physical tools, that answers the question
- A systematic discrepancy between the correct answer and the judged answer

Some of these illusions are compelling enough to fool even an acknowledged expert in probability theory. To illustrate, consider the following, deceivingly simple problem:

You meet a man on the street who has two children. He tells you that one of them is a girl. What are the chances that his other child is also a girl?

Most people (including some faculty members of the computer science department at UCLA) answer "one half" to this question, since the sex of one child should not have any impact on the sex of the other child. Presumably, they form an image of a child being born, while holding in mind the fact that another child was a girl. Since there are no known tendencies for order effects in maternity, the imagined birth should not be influenced by the sex of the other child and the answer therefore seems to be one half.

Now consider, however, how this problem would be handled using the mechanisms of formal probability. Families with two children, at least one of whom is a female, defines the sample space. In this sample space are three different non-overlapping subsets, families with two girls, families with a first-born girl and a second-born boy, and families with a first-born boy and a second-born girl. Now, granting that the sex of the first-born should not affect the sex of the second-born child, each of these three subsets should be equally likely. The probability of two girls is therefore one-third.

Problems of this kind have generated some still unresolved debates about the interpretation of subjective probabilities, judgments about causality, and the reasoning styles of normal individuals. The important question from an applied perspective, however, is relatively simple: do there exist real world contexts in which well-trained human decision makers systematically misjudge the true probability of a crucial event, decreasing thereby their ability to identify and communicate the risk factors involved in a particular situation?

Obviously, word problems like the one given above are not likely to compromise any of NASA's operations. Expert judgments about the probabilities of events and risks, on the other hand, are intricately involved in some, if not all, of the most important decisions taken by NASA engineers. Since these judgments cannot be codified and analyzed for objectivity with respect to an individual's knowledge and training experiences, the potential certainly exists for systematic biases to influence the decision making process.

One recent example of the devastating effects of human errors was the loss of the Mars Climate Orbiter in 1999. The mishap was eventually attributed to the failure of the navigation team to notice that the information provided them by outside engineers was in English rather than metric units. However, according to one official, the navigation team would have recognized the discrepancy if they had been more familiar with the Orbiter, if they had followed NASA procedures, or if they had communicated more effectively. In short, they made a number of decisions about how to approach the landing problem that ultimately compromised the mission, prior to issuing the misguided control instructions. Each of these decisions involved judgments about probabilities and risks, including the risks involved in skipping some procedures to save time and effort. If any single member of the team had decided that the model they were working with should be challenged, the incident may never have occurred.

2. The JPL Project

Judgment and decision making problems like the Mars Climate Orbiter occur across the spectrum of routine and mission-critical operations. Fully aware of this fact, NASA is interested in developing intelligent systems technologies to reduce the risks associated with human errors and to ensure the success of NASA's science and engineering activities. The primary objective of the HCC project is to develop a general method for human-computer design that can identify the situations conducive to human errors relating to probability judgments and provide feedback that can enable the operator to avoid them.

To do this, our first step is to introduce an adaptive feedback loop between operators and software that allows the user to be trained to "calibrate" his or her subjective probability judgments to the true risks inherent within a particular setting. The general concept is illustrated in Figure 1. The operator receives information from the interface which, in this example, includes both auditory and visual graphic data. Although there is an objective set of physical data presented by the audio and CRT display, Ω, the information actually obtained by the operator, γ, will depend on his/her sensory perception, attention, and many other factors. This internal mental impression of the objective information on the screen will form the basis for the operator's assessment of the situation, in this case, the current risk of

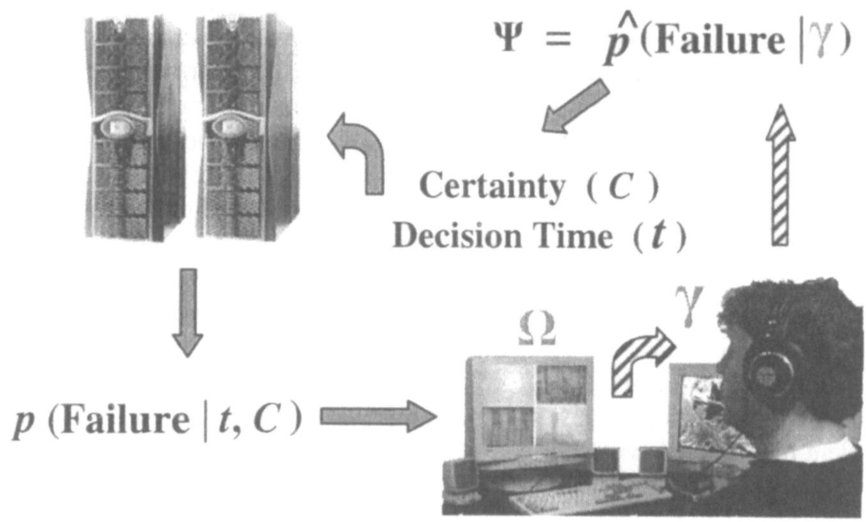

Figure 1. An illustration of the feedback loop between the information display and the operator's mental representation of the situation. The operator converts the display data to an impression about risks, which leads to a decision (Failure, No Failure), which is given with some degree of certainty (c) and at some point in time after the new information has been presented (t). The system computes the objective probability of a failure given this subjective report and provides feedback when these two conditional probabilities are miscalibrated.

system failure, $\Psi = \hat{p}(\text{Failure} \mid \gamma)$, where $\hat{p}$ denotes an estimated (and subjective) probability value.

The hatched arrows in the figure indicate a transformation from one information source (observable or unobservable) to an unobservable source. Solid arrows indicate that the information transmitted is observable. For example, the operator may be asked to report his or her certainty (C) in the imminence of a system failure. If the task requires continuous overt decisions about the status of system components, the time taken for the operator to decide may serve as a proxy for certainty, since there is usually a strong negative correlation between decision making certainty and decision time – uncertain decisions generally are slow and highly confident decisions are usually taken quickly [6-9].

If the operator's subjective probability state is "calibrated", it is equal to the true probability of a failure given the objective information presented on the screen,

$$\hat{p}(\text{Failure} \mid \gamma) = p(\text{Failure} \mid \Omega).$$

However, decision making biases and cognitive illusions might cause this subjective probability to be systematically miscalibrated. If so, then the operators' decisions may be suboptimal with respect to the goals they are pursuing. For instance, if the operator's task is to accurately identify the status of the system components (i.e., to maximize the overall proportion of times that good components are judged good and bad components are judged bad), the decision process would be suboptimal if the operator reports that a failure has occurred when the objective probability of a failure conditioned on the current psychological state, $p(\text{Failure} \mid \gamma)$, is less than one half.

Although γ is unobservable, a suboptimality in the operator's decision making strategy often can still be identified by substituting his or her observable subjective confidence reports (C) and judgment times (t) for the unobservable mental impressions that lead to these observable behaviors. That is, computing the probability of a failure conditioned on the operator's confidence report and decision time. It can be shown that if the operator's judgments are optimal with respect to γ, then they must also be optimal with respect to the observables, C and t. More specifically, if the decision strategy is optimal, then this implies that whenever a failure is reported with confidence level $C = c$, the objective probability of a failure conditioned on this report, $p(\text{Failure} \mid C = c)$, will be greater than or equal to one half.

Checking these conditional probabilities to see whether they are always greater than one half, therefore, provides a test for suboptimality of the operator's decision making style. Moreover, if a suboptimality is detected, providing feedback to operators by calculating $p(\text{Failure} \mid t, c)$ and informing them whenever this value is less than one half should afford them the opportunity to correct the suboptimality accordingly (i.e., by making the decision that is more likely to be correct).

In principle, as the description of the operator's behavior becomes more detailed (i.e., when it includes other information in addition to confidence and response time), the chances of detecting a suboptimality in an operator's decision making approach will increase. However, breaking down the operator's behavior into more and more subcategories decreases the sample sizes of the estimated conditional probabilities. An intelligent system must therefore examine a variety of possible descriptions and identify the ones that are most predictive of human errors in a given context. To do this, we use a search algorithm that covers various representations of the data and attempts to maximize predictive strength at a given sample size.

With minor variations, the same approach can be applied to situations in which a cost function is specified for the different decision making outcomes and the objective is to maximize the expected return. Experimental tests with novice and expert decision makers in our laboratory have shown that operators' suboptimal behaviors can in fact be identified from subjective confidence and decision time data and further that feedback can be used to correct these suboptimalities.

2.1 JPL/NASA Applications

Most of the critical decision making processes of interest to NASA cannot be easily reduced to a simple choice among a small set of possible actions. Usually the problems are open-ended, including the option, for example, to postpone the decision until more information can be obtained or other actions have been proposed and debated. In some contexts, such as the control of an aircraft or space vehicle, human behavior can affect the risk contingencies, increasing or decreasing them depending on physical actions at the controls.

To show how our general approach to interface design can be applied in these more complex settings, we developed a computer simulation that creates challenging problems of motor-control, speech recognition, causal reasoning, and risk assessment. Up to four pre-recorded speech sources consisting of flight-related information are presented simultaneously to the operator via headphone in a simulated 3-D scene. The operator may adjust the volume levels and spatial location of the sound sources by moving icons in a 2-D graph. Embedded randomly in the speech streams are a series of pre-defined keyword phrases that the pilot must recognize and acknowledge when they occur. Concurrently, the operator must monitor the positions of orbiting satellites and restricted areas that appear on a CRT screen, making decisions about the status of the satellites (e.g., when they are malfunctioning or off their intended paths) and whether their trajectories will put them at risk due to the vicinity of a restricted area.

The main control screen is shown in Figure 2. Using the mouse, the operator can redirect a satellite or manipulate its orbit. The events and risk probabilities change depending on how successful the operator is in controlling the movements of the satellites. Some of the keywords (e.g., "malfunction alert") indicate the need to make a decision about the components of the system. With experience, the operators learn some of the relationships and probabilities that are programmed into the simulation.

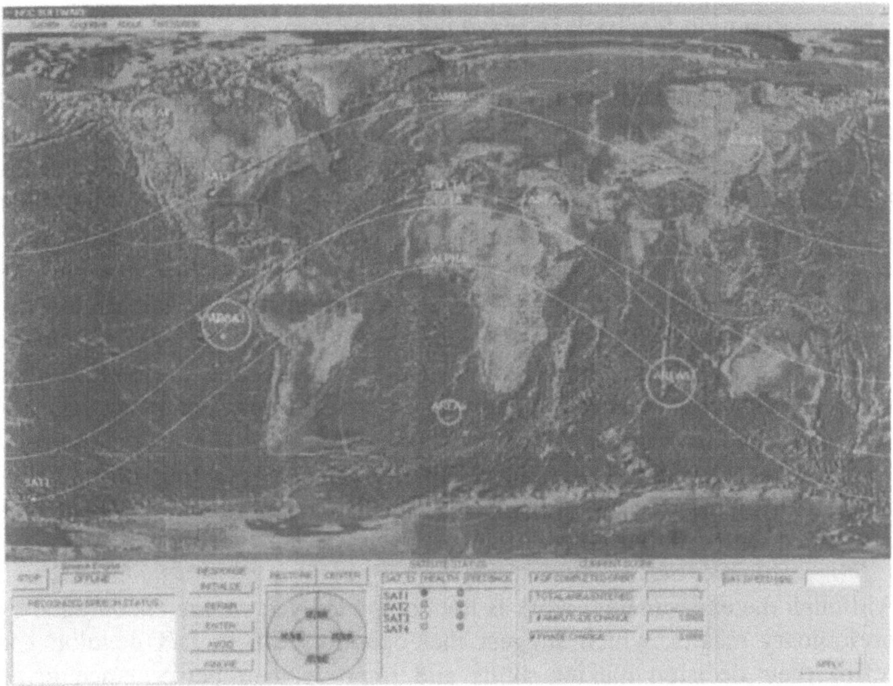

Figure 2. The main control screen in the satellite navigation simulation program. The operator monitors and adjusts the trajectories of up to four satellites while listening for warning sounds on up to four speech channels. If the satellites are allowed to enter a restricted area, they risk being destroyed, with this probability depending on the locations of other satellites and some other factors that the operator must learn to account for. The virtual locations of the audio channels can be adjusted by moving one of the four icons in the radial display at the bottom of the screen. Feedback about calibration is given on a separate screen that appears on command.

In addition to the feedback loop, several other components of the interface were adapted from previous work on human auditory perception and simulated 3-D spaces. Some of the key features include

- A user-configured, 3-D audio display with support for up to four external speech sources
- Automatic recognition of a pre-defined set of critical speech signals
- Capability to configure the simulation to involve a large number of interacting personnel.
- A GUI front-end to allow for quick and easy re-configuration of the problem scenario.

The re-calibration feedback is presented in the form of a vertical color strip with a rainbow pattern. Each point on the strip corresponds to a different degree of subjective confidence or decision time in a given decision. Green represents optimality and red represents suboptimality of the decision. Thus, an entirely green color strip for a given type of decision (e.g., "send repair team") indicates that for every level of subjective confidence reported when this particular decision was chosen, the objective probability that the decision was correct (e.g., a repair was in fact needed) was greater than one half. A strip that is green at the top and red at the bottom would indicate that for lower levels of confidence, the decision was suboptimal and should be replaced by a different one. That is, if the strip is red at the bottom, then when the operator feels it is necessary to send a repair team but has low confidence in this decision, he or she should choose the alternative course of action (e.g., ignore the warning). Intelligent agents acting on the data obtained from the operator actively seek for patterns of observable behaviors associated with suboptimal decisions.

Although the experimental work is not completed yet, we are encouraged by preliminary results, which suggest that operators can in fact develop and refine their decision making skills and improve their performance given proper feedback. In addition to our laboratory, investigators at the University of Florida have implemented this interface for use as a workload simulator and a basis for studying the decision making behaviors of operators under different degrees of stress and task demands. Other applications, including the communication and control signals exchanged between human operators and teleoperated robots, are currently being explored.

3. Summary

Over the course of the past few decades, judgment and decision making has become a major topic of research in the behavioral sciences. There are now literally thousands of published studies in this area, providing significant, and sometimes surprising, insights into the way both novice and experts reach decisions when they are faced with hard problems and have limited information to work with. During this same period, advances in computer and communication technology have been dramatic, making it possible to envision multimedia interfaces with ever-increasing speed, flexibility, complexity and intelligence.

The major premise of the HCC project is that interface designers must be quick to capitalize on developments in both of these disciplines. As a first step in realizing some of the potential for cross-fertilization, we have developed a general methodology that seeks to integrate established findings in the decision making literature with some of the recently maturing techniques of auditory, visual graphic and text display. The interface we have developed allows the user to configure and manipulate in real time a virtual 3-D audio scene, placing speech or warning sounds in the space by positioning icons on a 2-D graph. Automated speech recognition algorithms detect critical speech instructions and warnings, allowing the system to monitor the user's reaction to them. Concurrently, user-specific, intelligent feedback about the decision making style of an operator and its consequences allows the operator to recognize the presence of potential biases and cognitive illusions that would normally lead to suboptimal behaviors.

Our working hypothesis is that targeted, quantitative feedback identifying the precise situations in which operators' judgments are systematically misguided will reduce or eliminate the impact of these cognitive biases, despite the fact that they originate in intuitions that cannot be codified or communicated in objective terms. Preliminary data suggest that operators can in fact re-calibrate their intuitions, even in highly complex settings in which their actions affect the event probabilities and the feedback they receive adds an additional layer of complexity to the interface itself. In this respect, there is reason for optimism that intelligent, cognition-based models of the human-computer interface can be utilized in the future to prevent the occurrence of a large class of human errors, including some that are currently perceived to be inevitable consequences of human nature.

Acknowledgments

The authors were supported by NASA Ames Grant NAS7-1407.

References

[1] H.R. Arkes and K. R. Hammond (eds.). *Judgment and Decision Making*. Cambridge University Press, 1986.

[2] R. Hastie and R.M. Dawes. *Rational Choice in an Uncertain World.* Sage, 2001.

[3] W.M. Goldstein and R. M. Hogarth (eds.). *Research on Judgment and Decision Making*. Cambridge University Press, 1997.

[4] D. Kahneman, P. Slovic, and A. Tversky (eds.). *Judgment Under Uncertainty*. Cambridge University Press, 1982.

[5] M. Piattelli-Palmarini. *Inevitable Illusions.* John Wiley & Sons, 1994.

[6] J.V. Baranski and W.M. Petrusic. "Probing the Locus of Confidence Judgments: Experiments on the Time to Determine Confidence". *Journal of Experimental Psychology: Human Perception & Performance*, 24:929-945, 1998.

[7] D.S. Emmerich et al. "Response Latency, Confidence, and Rocs in Auditory Signal Detection". *Perception & Psychophysics*, 11:65-72, 1972.

[8] J.S. Shaw et al. "Recognition Instructions and Recognition Practice can Alter the Confidence-Response Time Relationship". *Journal of Applied Psychology*, 86: 93-103, 2001.

[9] D. Vickers et al. "Experimental Paradigms Emphasizing State or Process Limitations: II. Effects on Confidence". *Acta Psychologica*, 59:163-193,1985.

INDEX

MIX
Papier aus verantwortungsvollen Quellen
Paper from responsible sources
FSC® C105338

If you have any concerns about our products,
you can contact us on
ProductSafety@springernature.com

In case Publisher is established outside the EU,
the EU authorized representative is:
Springer Nature Customer Service Center GmbH
Europaplatz 3, 69115 Heidelberg, Germany

Printed by Libri Plureos GmbH
in Hamburg, Germany